Reputable Conduct

ETHICAL ISSUES IN POLICING AND CORRECTIONS

THIRD EDITION

John R. Jones, M.Ed., Ph.D.

PEARSON

Prentice
Hall

Toronto

National Library of Canada Cataloguing in Publication

Jones, John Raymond, 1943–
 Reputable conduct : ethical issues in policing and corrections / John R. Jones.—3rd ed.

Includes bibliographical references and index.
ISBN 0-13-123481-1

 1. Police ethics. 2. Correctional personnel—Professional ethics.
I. Title.

HV7924.J65 2005   174′.93632   C2004-900176-0

ISBN 0-13-123481-1

Vice President, Editorial Director: Michael J. Young
Acquisitions Editor: Ky Pruesse
Executive Marketing Manager: Judith Allen
Supervising Developmental Editor: Suzanne Schaan
Production Editor: Martin Tooke
Copy Editor: Trish O'Reilly
Proofreader: Heather Bean
Production Coordinator: Anita Heyna
Page Layout: Heidi Palfrey
Art Director: Julia Hall
Cover and Interior Design: Miguel Acevedo
Cover Image: Photodisc/Jacobs Stock Photography

34 35 36 CP 15 14 13

Printed and bound in Canada.

To my wife, Marilyn, and my three children, Steffan, Christopher, and Melissa, who have brought more love and meaning to my life than words can express. This book is also dedicated to the memory of my mother, Elizabeth Jones, whose generosity of spirit has left an indelible mark on my life.

Contents

Chapter 6   Subculture: What the Practitioners Think   72

Chapter 7   Tough Decisions   89

Chapter 8   Tools for Moral Decision-Making   106

Chapter 9   Where Do You Stand?   124

Chapter 10   Reputable Officers   155

Preface

This is the third edition of *Reputable Conduct: Ethical Issues in Policing and Corrections.* It is gratifying to know, judging by the response to the first and second editions, that the book has been read and discussed in many college classrooms and in training rooms of a wide variety of justice jurisdictions across North America. It is especially pleasing that an American edition of the book has been published and is now in its second edition.

Written for students in college justice programs and for police and correctional officers in training, *Reputable Conduct* has been designed as a working text rather than as a reference book. Much of the book has its roots in my own experience of teaching ethics for the past 20 years, and I thank my students for what they have taught me during that time. This experience has influenced the very personal tone of the writing style used in the book. As I wrote it, I imagined my students listening to me.

Reputable Conduct is intended to be a friendly and easy-to-read introduction to the difficult and sometimes peculiar ethical demands of the professions of policing and corrections. The book addresses some of the characteristics of these roles, with particular emphasis on subcultural constraints and how loyalty to our colleague group can sometimes cause us to sacrifice our individuality. How these constraints may affect the moral decision-making of an officer is examined in chapters 5, 6, and 7, which contain new material developed since the publication of the first edition.

One unusual feature of *Reputable Conduct* is the discussion in Chapter 4, which encourages the student to think about the role of the ethics educator. This discussion has been included to clarify the educator's role for the student. I hope it will reduce the possibility of misunderstanding arising between teacher and student regarding the purpose of an ethics course. I can't stress enough the importance of teacher and students engaging in this dialogue *prior to* embarking on any discussion of the rest of the book.

Chapter 2 was added after the first edition in response to a suggestion from some of the reviewers that some philosophical underpinning be provided. This chapter, somewhat expanded for the third edition, provides philosophical background and discussion on the concept of justice—given that this book has been written for those interested in the justice field.

Chapter 9 has been expanded to include several new moral dilemmas. Tools that may be helpful for resolving these moral dilemmas are discussed in Chapter 8.

The book concludes with Chapter 10, which addresses, among others, two important questions: 1) Should I admit a mistake when I make one? and 2) Is it possible to be an independent thinker *and* a team player? These issues may loom large in your thinking after reading and discussing the book.

Some of the material in *Reputable Conduct* could be described as sensitive—even controversial—in nature. To gain maximum benefit from the experience of working through the book, students are encouraged to consider with an open mind both the readings and the Reflections—searching questions interspersed throughout each chapter and based on the preceding content. *This will require a degree of maturity and discipline.*

Reputable Conduct is designed to provide students with a vehicle to promote both private thought and class discussion about issues that, from what I have been told by my students and practitioners over the years, are important to the vital roles played in our society by police and correctional officers.

If such thinking and discussion contributes, even in some small way, to the continuing service of justice, and to the individual officer's physical, psychological, emotional, and spiritual health, then the effort on all our parts will have been justified.

Acknowledgments

It has been a pleasure to work with staff members of Pearson Education Canada in the preparing of the third edition of this book. They have been unfailingly helpful, cooperative, and encouraging.

Particularly, I wish to acknowledge the early work of David Stover, Executive Editor. David was the person who first got me started on this book-writing adventure, and I owe him a debt of gratitude for his wise counsel and continuing support. His early suggestions for improving the first edition of the book were invaluable. Martin Tooke, Production Editor, shepherded the book through the production phase in a professional and timely manner and I wish to thank him for that. The expert help and advice of Trish O'Reilly is very much appreciated; her copy-editing skills enhanced the manuscript.

Finally, I wish to thank the many users of the first and second editions of the book for their encouraging and kind words about its usefulness. The book's success is due, in no small part, to the confidence you placed in it to assist you in making ethics an integral and important part of justice training. I do hope that the changes I have made in the third edition will further help you in that process.

J.R.J.

About the Author

Dr John Jones has worked for more than 30 years in the human development field as a minister, hospital chaplain, youth leader, correctional officer, high school teacher, probation officer, and correctional institution administrator. He has worked in Britain, Jamaica, Australia, the USA, and Canada.

For 21 years he worked as an administrator and professor in the Law and Justice Centre of a Canadian college of applied arts and technology, where he enjoyed teaching ethics to justice students.

In addition to his work as a researcher, John has been active in training and consulting in the justice field and in the corporate sector. He is considered a caring, compassionate, humorous, and skilled facilitator, seminar leader, and speaker, and has conducted seminars for thousands of participants in Ethics in the Workplace, Team Development with an Ethical Twist, and Hostage Survival Skills. He has also been called in court as an expert witness on the subject of officer subculture.

John holds a Joint Honours Bachelor of Arts degree in psychology and philosophy from the University of Wales, Cardiff; and a Master of Education degree and Doctor of Philosophy degree from the University of Toronto.

John is also the author of *Ethics for Career and Personal Success* published by Pearson Education Canada in 2001.

chapter one

Introduction

Perhaps the most important thing I learned from this class is that from now on I should always do the following:

- Keep an open mind about every issue.
- Try to look at an issue from as many perspectives as possible, before I form an opinion.
- Test my bias on issues about which I've formed an opinion.
- Realize that because everyone doesn't look at life's issues in the same way that I do, doesn't make them wrong and me right.
- I should not be ashamed to change my opinion on an issue when I had earlier felt the opposite.
- This class helped to make me aware of how grey many of life's issues are—there are few issues that are really black or white.

—Journal entry. Justice student in ethics course.

I think that an ethical person is someone who is consistent in doing the "right" thing. Being ethical consists of determining what is morally correct, and standing up for what you believe is right. I think an ethical person is a confident person. Someone who is not influenced by others from what is the morally correct action to be taken in a given situation.

I see myself as an ethical person but lack the confidence to stand up for every moral issue or ethical dilemma that I encounter.

—Journal entry. Justice student in ethics course.

CHAPTER OBJECTIVES

After reading this chapter, you should be able to

- Explain the purpose of this book
- Explain how we may have arrived at our values
- Explain how we in the justice field may inherit values
- Explain the notion of **subculture** in policing and corrections
- Apply knowledge of subculture in the military to policing and corrections
- Explain the notion of **de-individuation**

WHY THIS BOOK?

For 21 years I taught a course in ethics to students in the Law and Justice Centre of a Canadian college of applied arts and technology. For an equal number of years I have enquired of visiting publishers' representatives regarding texts available with respect to ethics education in the criminal justice field. The answer has always been that they know of little in this area. In addition, some of my faculty colleagues in other colleges, and some police and corrections training personnel, have asked me about such texts. On several occasions I was asked a similar question during a 1996 lecture tour of Australia.

That these enquiries are being made is encouraging, for there appears to be a growing interest in ethics training in the justice field. But the enquiries, along with some direct personal requests of me to do something concrete to meet the need, have translated into a personal challenge. You are now reading my response to that challenge.

WHAT KIND OF BOOK IS THIS?

This book is not intended to be a theoretical discourse on ethics. Background historical and theoretical information on ethics can be found, for example, in virtually any medical or business ethics text. This information is fairly consistently addressed in these texts, and I think it would be redundant to cover similar in-depth material here. Chapter 2 does take a cursory look at the theory and history of moral philosophy, but readers wanting a more in-depth treatment may wish to consult other sources.

This book is designed to be a practical work for instructors and practitioners in the justice field, and for their students. It is written and designed to be a resource in the classroom—both for the teacher delivering an ethics course, and most importantly for you, the student. My hope is that this book will become a *working* text—well-thumbed by the end of the educational experience for which it is being used.

Much of this work is characterized by an approach that I can only describe as deeply personal, and for this I beg the reader's indulgence. My experience gained as a correctional worker and institution administrator, before I became a college professor, serves as a foundation for much of what is discussed. And many of the ideas in the book spring from 20 years of trial and error on my part as a college professor working in a college justice department, as I have attempted to address ethical issues in my own courses. It has been both a privilege and a challenge to work with students over the years, and they have taught me a lot. The book's genesis and content have also been influenced by the experience I have gained in conducting ethics seminars with about 3 000 participants from various justice departments in Canada, the US, and Australia. Like my students, these participants have also contributed greatly to my learning. The work also has its roots in the research I completed for a PhD dissertation at the University of Toronto.

Briefly, the dissertation addressed ethics teaching in the justice field; it also examined the potential effects of officer subculture on the ability of individual officers to make independent moral decisions in their workplace. The dissertation employed a qualitative research methodology, which means that much of the data was gathered through in-depth interviews with practitioners from the justice field. Most, but not all, of these interviews were with correctional officers and administrators. Disappointingly, I experienced some difficulty obtaining cooperation from police forces for the purposes of interviewing their staff about this sensitive dissertation topic. Since completing my dissertation, however, I have had access to a number of police personnel who were prepared to be interviewed for this book.

Reputable Conduct, then, is clearly coloured by my own personal experience as a worker and administrator in corrections, by what I have learned from students and participants in my ethics courses and seminars, and from the research for my PhD dissertation.

THE INTENT OF THIS TEXT

Though meant to be thought-provoking, this text is not designed to condemn or accuse or point a finger. The roles of the correctional officer and the police officer are often demanding and extraordinarily complex ones. It would be improper, therefore, for this text, or any similar text for that matter, to peddle a particular point of view with respect to the complex ethical issues that often emerge from this kind of work. There may be times when you feel, however, because of the nature of the challenges presented and the questions asked, that there is a hidden agenda—that is, that there is a preferred way of thinking about a given issue and that you are being led in that direction. Because students can sometimes feel this way, there is a discussion in Chapter 4 about what I call the lurking dangers of indoctrination, or the perception by students of indoctrination by the teacher. The intention of this book is not to lead you in any particular direction or to tell you what you should do in a given situation. The intention is to prod, push, and poke you into thinking more deeply about what are, it would seem to me, important ethical issues and moral dilemmas facing those of us involved in the justice field.

If the discussions and dilemmas are on occasion posed in a manner that causes you to feel that you are being judged, this is because they are designed to be searching and challenging. It is not my intention to stand in judgment over you, the reader. On many occasions, I have personally confronted similar dilemmas in my own career as a correctional officer and as an institution administrator, and only a foolish person would claim infallibility in dealing with such dilemmas. Indeed, I am still not sure, many years later, if I did the right thing in some of the situations I faced. Because of my own struggles, and the struggles of others who have conveyed their stories to me, my approach is, I hope, one characterized by compassion. But this does not mean that we should shy away from asking each other some hard questions about our work. As is the case with almost any field of ethical enquiry, there are many more questions than there are answers, so throughout the text we will be stopping, under the heading Reflections, to ask some of these questions.

Perhaps an appropriate first challenge might be for you to debate whether talking about moral dilemmas facing us in our daily work is a worthwhile activity.

Reflections

Might we learn something from considering alternative perspectives of colleagues, subordinates or superiors?

Is there anyone who has a monopoly on wisdom with respect to the dilemmas we face daily?

Should we go it alone with these dilemmas—that is, work out our own solutions in isolation from our colleagues?

Should we find out what our peers think about specific issues and follow the consensus?

Where does talking about moral dilemmas and their resolution rate as a teaching priority? What about moral, as well as legal, personal accountability?

Does moral accountability matter enough to us to warrant serious discussion about serious issues?

Can this difficult job we do cause our sense of what is right or wrong to become confused?

Can our moral sense be changed, or sensitized, or perhaps reawakened?

These are, in my view, examples of important questions deserving serious attention. Perhaps you will consider this point of view itself worthy of discussion. If you disagree with me, then you should be able to give reasons for your disagreement. In any event, the above examples are the sort of questions that will be asked from time to time throughout the book. It would be an interesting exercise—at the *end* of the book and your course—for you to reflect on whether you would make any changes in the way you earlier attempted to answer the questions.

IS THERE A LINK BETWEEN ETHICS EDUCATION AND MORAL BEHAVIOUR?

This text tries to provoke discussion because I believe that conversation about moral dilemmas and their resolutions has the potential to heighten awareness of personal moral responsibility. Derek Bok (1988), while president of Harvard University, wrote a thought-provoking article called "Can Higher Education Foster Higher Morals?" Bok has this to say:

Skeptics will reply that courses in moral reasoning have no effect on behaviour, but this criticism seems overdrawn. To be sure, no instruction can suffice to turn a scoundrel into a virtuous human being. But most young people arrive at the university with decent instincts and a genuine concern for others. For them, courses that foster an ability to detect ethical issues more perceptively, to think about them more carefully, to understand more clearly the reasons for acting morally seem likely not only to train their minds but to have a positive effect on their behaviour as well (p. 7).

Of course, Bok's notion applies equally to those of you in college justice departments, and in police and corrections training facilities.

I tend to agree with Bok. We shouldn't have unrealistic expectations for the effects of ethics courses on people's behaviour. The potential is there, however, for informed and critical reflection and thought to translate into a greater desire to do what one considers right. Many participants who have attended my ethics seminars have said things to me that seem to support this idea. Talking about ethics, about what may be right or wrong, certainly raises our level of consciousness about important issues. And being more aware, more thoughtful about the dilemmas we face can sometimes translate into more courageous decision-making and actions. But what I think about this issue is not important. What matters most is what you think.

Reflections

What do you think of Bok's point of view?

Do you believe that thinking more clearly about ethical issues and moral dilemmas has any potential for helping you do what you consider to be the right thing?

THE PLACE OF COURAGE IN ETHICS EDUCATION

Taking time out in a course like this to think about ethical issues may, as I have indicated, foster moral courage—that is, having worked out what we should do in a given situation, we may find the strength to do it, irrespective of the forces pulling us in other directions.

But the first sign of courage, perhaps, is for each of us to be willing to face up to the hard questions posed earlier, and others like them, in this text. To risk the intellectual, psychological and emotional discomfort that will invariably result from such honest enquiry is a bold step. It can be disturbing to find that the result of such enquiry is simply, on occasion, little more than the replacing of answers we thought we had with troubling questions concerning those "answers." It can be equally disturbing to be confronted with the thought that if we are to be honest with ourselves, our particular way of looking at something, or a particular value we have held for some time, may need revision—if not complete re-thinking.

Students often talk about being somewhat scared of confronting their biases and prejudices. The good news is that those who show the willingness and the courage to submit their values and ideas to scrutiny often feel they are stronger for the experience. One of my students, referring to his personal growth, wrote at the end of an ethics course, "Not only does this course challenge your existing values and norms, I found myself asking why do

I think like this? and what causes me to think like this?... I think if you are comfortable with this self-examination, that it opens up doors for self-growth." Another wrote, "We were afraid and we grew.... I have realized that being afraid is normal." And another, "This has been a challenging class. While I have often left the room with a headache it has certainly made up for it in other ways. True morals are those that can be challenged openly and objectively. If a person can do that and still believe in that moral, then it is certainly a moral that should be believed in."

Reflections

Is being willing to submit your views and values to a significant challenge the kind of activity you consider requires courage?

Or, is it a foolhardy exercise, at best likely only to reduce your general happiness, at worst likely to confuse you totally?

There are people who would argue, and possibly with some justification, that unless we are able and willing to test our views and our values by the fire of honest enquiry and soul-searching, a) the values are not truly our own, and therefore, b) they are not worth having. Many of our views and values, it is argued, are inherited–put on, as it were, as we would a suit of clothes. We have just assumed them, often without paying much attention or thought to how we arrived at them. If this is the case, then perhaps we should call them habits rather than values. What do you think?

INHERITING VALUES IN THE JUSTICE FIELD

Some would argue that the tendency toward inheriting views and values described in the Reflections segment above applies particularly to those of us who work in the justice field. Some would take it a step farther and argue that to resist inheriting or adopting the values of our colleague group is a decision fraught with some risk. These values are what we sometimes call *norms*. For example, Frederick Desroches (1986), a sociology professor at a Canadian university, speaking of new recruits to police work, makes this point:

> From association with fellow police officers, and by being sensitive to their attitudes and beliefs, the new recruit soon picks up on the general group's opinion about most police matters.... Once a subculture has been formed, novices entering the group are confronted with a "fait accompli"—a set of norms and values to which they are exposed and are expected to conform (p. 40).

And Elizabeth Grossi and Bruce Berg (1991), in the corrections context, make a similar point:

> Again, it may be that in order to attain a satisfactory level of peer support, and reduce conflicting pressures endemic to correctional institutions, correctional officers must compromise their personal values and interests (p. 80).

These are sobering and troubling sentiments. The writers indicate that to survive as a police or correctional officer in terms of early and continuing acceptance by peers, an individual officer would have difficulty operating as an independent person, doing his own thinking about moral dilemmas encountered on the job. Desroches talks about police officers, especially new ones, having to conform to group norms and values. Grossi and Berg, in talking about correctional officers, go further and talk about the requirement to compromise personal values and interests in order to be accepted by peers.

Later we will discuss this issue of colleague loyalty—or peer pressure, if you like—in greater depth. The issue has to do with **subculture** in policing and corrections, and how it can impact on individual officers.

Reflections

What do you think of the views of these researchers and writers? Do they have any merit? Or do these two authors have it wrong?

A recent high-profile case from Canada's armed forces may shed more light on this troubling question.

SUBCULTURE IN THE MILITARY

In 1993, an event unfolded—and I hope I am not being too melodramatic here—that scarred the soul of Canada. Consideration of that event here may help to provide us with a useful context for discussions later in the book. Historically, Canada has taken some pride in the fact that it has contributed to world order and peace by deploying its troops to many parts of the world for peacekeeping duties. Canadian soldiers have been characterized as something approaching angels of mercy and have, more often than not, been honoured and accepted by the citizens of countries in which they were serving.

In 1992–93, Canadian troops were sent to Somalia on a peacekeeping mission. The soldiers were members of the Canadian Airborne Regiment. A subsequent inquiry has discovered that some of these peacekeeping soldiers conducted themselves in a brutal, racist manner, and, in fact, tortured and murdered a 16-year-old Somali youth. One disturbing discovery that has emerged from the inquiry is that the military has its own entrenched subculture, and that this subculture is often characterized by some of the things Desroches and Grossi and Berg talk about. One of the chief characteristics is the tendency for recruits to allow themselves to "buy into" the subculture in such a way that their loyalty to colleagues can transcend the responsibility they have to think for themselves and to act according to their own moral sense. Indeed, one military police officer admitted that "his investigations were thwarted by soldiers covering up for each other." While giving evidence at the Somali inquiry, this military police officer added, "We're soldiers first and tradesmen second," later suggesting that if he were the one being questioned by the police, he would probably have kept quiet as well (*Toronto Star*, 15 October 1996, p. A2).

In 1993, on the CBC radio program *Morningside*, during a discussion about the Somali affair between host Peter Gzowski and a retired lieutenant-colonel of the Canadian military, the lieutenant-colonel referred to what she described as the "dangerous bonding" that can take place in military units; this bonding, she claimed, effectively inhibits—if not *prohibits*—individual moral decision-making. She asserted that the individual (soldier in this case) must constantly be reminded that he continues to carry full responsibility for his moral decision-making, irrespective of the prevailing ethos or subculture.

DE-INDIVIDUATION

It may be useful to contrast the lieutenant-colonel's point of view with something psychologists call **de-individuation**. Psychology professor James Ogloff defines de-individuation thus: "There is pressure exerted by the group and people lose the ability to regulate their individual behaviour." He adds, "You are not held accountable for what has gone on, so you end up doing things you wouldn't do individually" (*Maclean's*, 30 January 1995, p. 18).

It may be difficult, however, to reconcile the lieutenant-colonel's view that individuals continue to carry full responsibility for their actions when, if Ogloff is correct, peer pressure or subcultural constraints can cause individuals to lose the ability to regulate their behaviour. In other words, in a sense they become victims. (This apparent conundrum would make a lively topic for class debate.)

A Confession

In a haunting article entitled "Discretion No Part of Valour" in the *Globe and Mail* (1 October 1996, p. A6), Eron Main, a former Canadian military officer, refers to the Somalia affair. The article is a confession of sorts. Main writes, "The unofficial military code I lived by involved turning a blind eye to my fellow officers' misdeeds. Ultimately such 'loyalty' led to the Somali inquiry as well as to my retirement." He adds, "We learned the lessons of loyalty a little too well, until we gave it where it was no longer deserved: to the officers who had debased their commissions, and were no longer good and honest." Main, in reflecting on his days as a student at the Royal Military College of Canada, talks about the code of loyalty: "In four years I had encountered many future officers who neglected the college motto—Truth, Duty, Valour—in favour of the unofficial code: Don't Get Caught. I knew of several whom I considered unworthy of their commissions. Of course, I would say nothing. We had been taught from our first day as cadets that we should never, ever, betray our friends, our classmates, our fellow officers."

The US Naval Academy is an historic and tradition-bound institution charged with turning out leaders of the US military. One thing that has always made that school (and others like it) seem to stand above and apart from the rest has been its stated support for, and adherence to, a code of honour. That code essentially maintains that (and I am paraphrasing here) "we will not lie, cheat, or steal, nor tolerate those who do." Several years ago, when the Naval Academy became involved in a series of scandals including academic cheating, drug sales and usage, and auto thefts, the television program *60 Minutes* conducted a series of interviews. They learned that in addition to the formal Code of Honour there was another less formal but in many ways more powerful code that held "Never bilge your buddies." In the police and corrections arena, that code would caution practitioners to never inform on or "rat out" their peers.

Another way in which we characterize this code is to refer to it as a cult or code of silence.

An internal study of the Canadian Armed Forces, written by Captain Claude Beauregard, was presented to his superiors in the spring of 2000. Captain Beauregard, perhaps courageously, claimed in the report that the Canadian Armed Forces had not progressed very far from the way they were conducting business at the time of the Somalia affair. The military, he claimed, was still given to stifling accountability, and the problem was a systemic one. He described the root of the problem as "a military culture based on a code of silence."

These sentiments echo many of the thoughts and comments expressed to me by respondents in my dissertation research. They also echo many of the comments made to faculty in our justice department by students returning to classes after serving their field practica with police forces or in correctional institutions. Now let's try to relate what we have been discussing to justice students and the college classroom.

Knowing, Yet Not Knowing

One of the more difficult challenges I had to face in my teaching career was dealing with a group of angry justice students who had decided to punish a classmate. The classmate had expressed dismay to a teacher that a number of students had cheated on an examination. The teacher, after handing out an examination paper and instructing the students to leave the papers upside down on their desks, realized she was two papers short. She told the students to leave the papers upside down and went back to her office to photocopy some extras. While she was gone, it appears there was open season on exchange of information by approximately two-thirds of the students! Several minutes later, the teacher returned to a now-quiet classroom and instructed the group to turn over their papers and commence writing the examination. One by one the students finished writing the examination and excused themselves from the classroom. The last student to leave was the one I mentioned above. She said to the teacher, thinking there was no other student within earshot, "I have a problem with your leaving the classroom. It allowed several students an opportunity to cheat, and I don't feel that is fair." One of the other students was waiting outside the door and overheard the comment. What happened in the next few days was alarming but may come as little or no surprise to many of you reading this account.

Word got out and the reporting student was treated mercilessly by her formerly friendly classmates. On the second day after the incident, she came into my office in a fragile state to declare she could not put up with the hostility anymore. Even though she was in the final term of a two-year program, she had decided to drop out of school. She had committed the irrevocable mistake of betraying her classmates. (Perhaps it was Eron Main's fear of this kind of reprisal that caused him and his military colleagues to say nothing about the major indiscretions of classmates and colleagues.)

The most sobering thought about this event is that these were the same justice students who, the previous semester when we were discussing the notion of subculture in a class, had protested that there would be little or no chance of their ostracizing a colleague for doing the right thing. As discomfiting as the examination incident was, and as painful for the student as it was, it provided a useful vehicle for discussion and learning. The students and I had a three-hour discussion that helped to clear the air and to clarify some values. The points of view of both "sides" were heard, and, as a result, cooler, and perhaps wiser, heads prevailed. The honest student stayed in college, graduated, and is doing an excellent job in the corrections field.

> ## Reflections
>
> What do you think about values and attitudes that can be picked up by new recruits to the military, to policing, or to corrections?
>
> Should loyalty to your colleagues be fostered as an appropriate and, indeed, necessary value? If so, what consequences might inheriting this value have on your ability to make the decisions you—and you alone—feel you must make in the context of your work?

SUMMARY

The primary purpose of this book is to stimulate you, the student, to think courageously and deeply about ethical dilemmas. The issue of colleague loyalty above all else is just one of many that seem to be an integral part of working in the justice field. To be able and willing to engage in honest, open discussion of issues like this one demands a high degree of maturity and courage. Resisting the temptation to react defensively is a real challenge. Of course, if you consider the ideas and questions in this book in a mature and honest way, and you are still convinced that your original beliefs can be justified, that is fine. The intention is not necessarily to change your thinking, though that may happen on some issues; it is simply to provide you with an opportunity for considering the issues in a deeper way.

The idea that it is often difficult to remain one's own person in the face of subcultural pressures is a central theme of this book.

Another important purpose of the book is to encourage you to work toward personal resolution regarding some of the issues raised, where possible. To achieve this you will, of course, need some tools to assist in the process. These tools are available and will be discussed in detail in Chapter 8. The book will also provide opportunities for you to practise using these tools as you struggle to reach conclusions about ethical dilemmas.

The dilemmas offered for discussion are not hypothetical. It is my hope that they will touch a chord in you because of their practical significance. They are real. They have been experienced and recounted to me either by my students while on their field practica, or by officers and administrators of police forces and correctional institutions in Canada, the US and Australia. Some of the specific details have been changed in the interests of ensuring that no police force, correctional institution, or individual is identifiable from the information given. Some of the dilemmas are ones more likely to be experienced by police and correctional institution administrators, and others by frontline officers. I would encourage you to try thinking your way through both the administrator and officer dilemmas. In this way, you may become more sensitized to the dilemmas others experience.

Reflections

Should a police force or correctional institution manager encourage subordinate officers to think for themselves?

How do you feel about this idea of thinking for yourself, especially at times when your colleagues may be suggesting strongly that you not do so?

These Are Tough Questions

Here's a little teaser for you.

We live in unpredictable and, some would say, dangerous times. Osama bin Laden and his followers have made us all perhaps a little hypersensitive. As I write this piece in the summer of 2003, there have been several media reports indicating that some devout American Muslims have been taken into custody and are being held indefinitely. No charges have been laid against them. One is a Muslim cleric from Ann Arbor, Michigan, who was taken into custody at a federal prison in Chicago and has been seen only infrequently by his family since. Presumably the FBI or the CIA have some suspicions about this person and other individuals but have no firm grounds on which to lay charges.

Now we may ask ourselves: Is the indefinite incarceration of these persons fair? Remember they have not been charged with the commission of any crime. Is this state of affairs a just one? Is it appropriate to suspend the rights of a few individuals in order to protect the interests and well-being of the majority? Try not to jump to a conclusion about these questions. Think hard about them and consider all the angles.

Let's take the situation one step further. What if there are very strong suspicions that one of these incarcerated individuals is implicated in the events of September 11, 2001 and was plotting other attacks on innocent citizens? There is no proof, but there are grounds for concern. *Do you think that this would justify the use of torture on this individual?* It's clear that if other attacks are imminent it places a tremendous burden on the government to discover what those attacks might be and where they might take place.

Imagine, hypothetically, you were present at an interrogation and torture session of one of these individuals whom, it is conjectured, *may* be linked to the Taliban. Would you feel comfortable with the notion that an American citizen, (or Canadian citizens for that matter because the same thing is happening in Canada) against whom nothing has been proved, is being pressed for information in this way? In other words, is a little bit of wrongdoing justifiable (if, in fact, torture is a wrongdoing) if the end result is greater safety and security for the general population?

Northwest Airlines has made a decision to pay damages to one of its passengers. On 25 December 2001 a Pakistani man from California, Harris Khan, was asked to leave a plane bound for San Jose from Minneapolis because of his "Middle-Eastern appearance." As part of the agreement the airline has issued an apology to Mr Khan.

There are many who feel that this kind of action amounts to religious and/or racial *profiling*, and is an over-reaction to our safety and security concerns. Not only is it an over-reaction, it is claimed, but it is unjust and unfair.

Edward L. Greenspan, one of the top Canadian lawyers, wrote a piece about this issue in *Maclean's* magazine. "There is a land where the government can arrest non-citizens," he writes, "throw them in jail for an indefinite period of time, and then remove them from the courts, all in virtual secrecy.... No, this is not China or Cuba.... This is Canada—a country I believed understood the transcendent importance of safeguarding civil liberties." He adds "The challenge is to figure out a way to deal with the threat of terrorism without losing the freedoms that make Canada the great nation it is." He quotes Earl Warren, former Chief Justice of the US Supreme Court over 35 years ago: "It would indeed be ironic if, in the name of national defense, we would sanction the subversion of one of those liberties which make the defense of our nation worthwhile."

Have you thought about this important issue?

While we are on the subject of profiling, a very hot issue in policing these days is racial profiling. Let's think of this difficult issue for a few moments.

Racial Profiling

This is a topic that is exciting a fair amount of interest from the media, and a vociferous reaction from police departments, primarily of vehement denial that they practise it.

Racial profiling is said to exist when a member of a visible minority group is stopped and questioned by a police officer *simply on the basis of race*. So if a rather untidy Black person, let's say, is walking normally along a street carrying an obviously expensive briefcase, an officer who stops and questions such a person would be guilty of racial profiling. What may drive the intervention are the following assumptions: Much crime is committed by Black persons. This person is Black. It seems incongruous for such a person to own and carry such an expensive briefcase. Therefore, this is a situation that calls for the person to be questioned because the briefcase may well have been stolen.

Of course, the problem is that if the person were White, the intervention may very well not have occurred. The intervention, therefore, was done primarily on the basis of the person's colour.

Well, we might say, the person was untidy and there *is* some incongruity in such a person carrying an expensive briefcase. Perhaps. But do you know any journalists, for example? Many of them who work for large newspapers or magazines are very well paid and can easily afford expensive equipment like a briefcase. Equally, some (many?) of them present as very scruffy, unkempt individuals indeed. (We have a son who illustrates this. He is paid handsomely for working as a columnist for an international, high-status magazine but you certainly wouldn't know it to look at him!)

He has never been stopped on the street and questioned. But then he's White. The question has to be asked: might his experience be any different if he were Black?

In the spring of 2003 a Toronto Raptor basketball player was acquitted of a drunk-driving conviction after the judge ruled he had been pulled over by the investigating officer because of his colour. Because of his status as a highly paid sports star he owned a high-end sports utility vehicle, which he was driving on the night of his arrest. The officer responded to the perceived incongruity between colour and class (represented by the expensive vehicle) and thought the vehicle might have been stolen. He pulled the vehicle over on that basis, not because it was being driven erratically. There were no reasonable and probable grounds; the driver just happened to be found to be intoxicated. The officer had no suspicion the driver was intoxicated prior to questioning him. The question has to be asked: Would the vehicle have been pulled over had the driver been Caucasian? If the answer is a probable no, then would it be reasonable to assume there was something else at play here?

At roughly the same time as this event, the *Toronto Star* released a research report it had prepared on the topic of racial profiling by police in the Greater Toronto Area. The report concluded that there was strong evidence to suggest racial profiling does exist and that the research methodology and the resulting conclusions had been corroborated by several well-qualified, independent researchers.

In response to this situation, there has been a vehement denial from the Metropolitan Toronto Chief of Police, from the Police Services Board, and from the Officer Association that racial profiling is practised by police officers in this force.

Sadly, we live in dangerous days. "Terrorism" is a word we seem to use far more often than we used to, and for good reason. As I write this, several Westerners living in a compound in Saudi Arabia have just lost their lives after a terrorist car bomb exploded in their compound. Those of you who have recently crossed borders or taken a flight will have noticed the level of surveillance has been raised significantly. It is not an unusual experience any more for passengers at airports to be asked to remove their shoes for inspection. Who, even five years ago, would have thought this was a practice we would simply come to resign ourselves to? We are questioned about our luggage, and it and we are processed through myriad screening devices. All for our own safety, of course. So we shrug our shoulders and accept it as part of our existence in this modern world.

But what if we are dark-skinned? What if we have a swarthy complexion? What if, by our dress and complexion, we announce we may be Muslim? We know that Muslims are generally law-abiding, industrious, productive members of our society. We also know that in the current political climate terrorists are often Muslim. These terrorists, however, are invariably drawn from the extremist, fundamentalist ranks of that faith, whose activities and ideology are abhorrent to Muslims and non-Muslims alike.

So what is a security person or a customs officer to do when screening someone who "looks like a terrorist might look"? Should these individuals be singled out for extra attention? And, if so, on what basis? Is it simply on the basis of the way they are dressed or the colour of their skin?

I spoke at a police ethics conference in the US at which one of the people attending was an officer from a police department in British Columbia. He was an East Indian man with a swarthy complexion. The conference was held a short time after the terrible 9/11 bombing in New York, and security at all airports and borders was extremely high. Two of this man's colleagues were also attending the conference, and they described to me how their colleague had been singled out for intensive scrutiny at the airport. It was only after they had vouched for him and had produced their police identification that he was allowed through. Apparently, his own police identification was being regarded with suspicion. Here's at least one police officer who would have some understanding of how Mr Khan must have felt when he was asked to exit the Northwest Airlines flight.

It's tough to know how we should respond, isn't it? On the one hand terrible things happen because screening fails, and, on the other hand, if we are too cautious and/or aggressive we may become discriminatory.

Perhaps we would have a clearer view of the situation if we ourselves were subjected to extra attention on the basis of how we look.

Reflections

Is it possible that racial profiling, to a lesser or greater extent, exists?

Can a police force be said to be guilty of racial profiling even if it does not have an official policy requiring its officers to give special attention to certain minority groups when on patrol?

If you are a Caucasian student or police officer, have you ever asked a Black colleague for his/her opinion on this issue?

One final note: The fact that the situations included for discussion are *dilemmas* indicates that they have no hard and fast solutions. Indeed, there may well not be one right answer. What we can do is make a choice that is the best available, while acknowledging that the perfect solution does not exist.

Before we get to the point in the text where these dilemmas are posed, however, you will see that chapters 3 and 4 address important topics which, I hope, will become subjects of debate and discussion almost before anything else is discussed in your class. Chapter 3 addresses the ethics of teaching ethics in justice programs and Chapter 4, as I mentioned earlier in this chapter, addresses the role of the ethics educator with particular attention to the possible dangers of indoctrination. In my experience, it is unusual to encourage students, either in a book like this one or in an ethics course, to participate in a dialogue about these issues. However, discussion of these topics with your instructor can help to set the tone of your classroom experience. It may also help to establish some parameters.

More importantly, perhaps, it may save you and your instructor some of the pain and anguish my ethics students and I have experienced in the past over fuzzy role expectations, unclear teacher intentions, and miscommunication regarding the purpose of the course.

chapter two

Ideas

Man is not to blame for what he is. He didn't make himself. He has no control over himself.

—Mark Twain, novelist

Nature is strong and she is pitiless. She works in her own mysterious way, and we are victims. We have not much to do with it ourselves. Nature takes the job in hand, and we play our parts.

—Clarence Darrow, US defence counsel

We are responsible.... It is not true that everything that happens to us is like "being struck down by a dreaded disease."

—Sidney Hook, philosopher

He who lets the world, or his own portion of it, choose his plan of life for him has no need of any other faculty than the ape-like one of imitation. He who chooses his plan for himself uses all his faculties.

—John Stuart Mill, philosopher

Man has free choice, or otherwise counsels, exhortations, commands, prohibitions, rewards and punishments would be in vain.

—St. Thomas Aquinas, theologian

Give me a dozen healthy infants ... and I'll guarantee to take any one at random and train him to become any type of specialist I might select—doctor, lawyer, and yes, even beggar-man and thief, regardless of his talents, penchants, tendencies, abilities.

—John B. Watson, behaviourist

CHAPTER OBJECTIVES

After reading this chapter, you should be able to

- Define moral philosophy and "philosophizing"
- Explain the potential dangers of gullibility and mindlessness
- Explain **Dr Martin Luther King, Jr**'s ideas about good laws and bad laws
- Explain the difference between morals and ethics
- Explain what a moral dilemma is
- Explain what we might mean by the term "a wise person"
- Define **libertarianism** and **determinism**
- Explain **utilitarianism** and identify two major writers who developed the concept
- Explain the **principle of equal respect**
- List and explain some ideas about justice developed by **Plato, Socrates, Aristotle, Hobbes, Kant,** and **Sartre**
- Explain Plato's idea that how we behave is closely linked to the thinking we do about our behaviour
- Explain Socrates's ideas about teaching (the Socratic Method)
- Explain Aristotle's contribution to philosophical thought regarding logic
- Explain what Hobbes meant by "social contract"
- Explain Sartre's view that "man is condemned to be free"

INTRODUCTION

This is a foundation chapter, so to speak, for the rest of the book. It provides some ideas that a few selected thinkers and writers have had about society in general, and about justice in particular. Try to keep these ideas and concepts in mind when you move on through the book. Make whatever connections you can, and never hesitate to refer back to the ideas discussed here, because they may help to shed some light on dilemmas raised later.

It's always a pleasure for teachers—and I very much include myself here—when they see their students making connections with ideas they have previously learned. It means a) at least some of the things the students learned previously have stayed with them, and b) the things learned are not locked away in some kind of mental "compartment." The best students are the ones who can relate ideas and concepts from one class to another, and from one course to another. Let me encourage you to engage in the kind of thinking that draws from any and all of your educational and life experiences.

Let me further suggest to you that learning done in separated, isolated, digestible pieces for later regurgitation in tests or exams is, in all probability, not the best kind of learning, because it relies essentially on remembering, parroting back, and then forgetting. Do yourselves and your teachers a favour: look at the ideas and discussion in this chapter as part of an integrated whole.

"Keep It Simple Stupid," they said. Now there's a challenge.

Before your eyes glaze over, hang in there! I am going to attempt to make this background information as interesting and as relevant as I can.

First, we will address some common recurring questions: What is moral philosophy? What does it mean to philosophize? What's the difference between morals and ethics? What is the nature of a moral dilemma? What is a "wise person"? Second, together we will discuss some ideas that certain important people have had about moral philosophy. We will do so *in a context* in an attempt to make these ideas come to life. Third, we will examine some ideas of just a few of the major players in philosophical thought. We will take a look at three ancient Greek philosophers, Plato, Socrates, and Aristotle; one seventeenth-century philosopher, Thomas Hobbes; one eighteenth-century philosopher, Immanuel Kant; and one contemporary philosopher, Jean-Paul Sartre. Of the many big contributors to philosophical thought, these thinkers are the ones I have chosen because of some of the observations they make about the human condition, and about justice specifically.

One of the biggest challenges for any writer of a chapter like this is to convey the material in a meaningful way. The risk as I attempt to do so, is that I will oversimplify, and in the process do a disservice to those thoughtful, clever, wise and insightful philosophers. I hope they will forgive me—all of them will have to do so from their graves!—for any distortion of their complex ideas, as these ideas are examined through my particular prism in an attempt to be interesting and to "keep it simple."

WHAT IS MORAL PHILOSOPHY?

The word *philosophy* simply means "the love of wisdom." The word has its roots in two Greek words: *philein*, which means "to love," and *sophia*, which means "wisdom." The purpose of philosophy is to explore important matters concerning the world and the universe and those of us who live in it. Moral philosophy, or ethics, is one of several different philosophical disciplines and it searches for knowledge and meaning in the areas of human behaviour and morality. Philosophers engage in an ongoing search for truth and understanding, and they do so more by asking important questions than by attempting to find and provide prescriptive answers to life's mysteries. So you would never find a moral philosopher, or any other kind of philosopher, for that matter, who would feel comfortable telling you as an individual how you should behave. That search for truth and for a sense of what may be right or wrong with respect to a certain action is *your* search. And the more you know about moral philosophy and what various thinkers have said about important issues, the more skilled you will become at thinking through your own situations.

WHAT IS "PHILOSOPHIZING"?

When you think about important questions, such as the rights of animals, the rights of the mentally challenged members of our society, abortion, capital punishment, social injustice, the meaning of life, and so on, you are philosophizing. You are searching for truth and meaning. The only difference between you and philosophers *per se* is that they are probably philosophers by profession; that is, they probably write about it and teach it. They will probably be more skilled because they have spent a lifetime studying the ideas of many great minds and have been trained in the rules of logic, thinking, and argument. That does not mean, however, that when we think deeply about important issues we are not philosophizing, too. So there is no magic to it, and we should never feel excluded from the process.

What is important is that we do our own thinking! To simply choose to go along with the crowd, or accept the status quo is to do ourselves a disservice. Not only that, but it's dangerous. We can be led up garden paths and end up deeply regretting our gullibility.

Many years ago in Britain a credible and well-respected news program called *Panorama*, hosted by a reputable and trusted broadcaster, hoodwinked almost the entire nation one April Fool's Day. The program aired a serious-sounding documentary on "The Spaghetti Trees of Northern Italy." The film footage included scenes of many acres of cultivated trees bearing their spaghetti harvest! The frightening thing is that viewers were prepared to suspend their own knowledge and beliefs because the information was being conveyed in a serious manner by what they considered to be a trustworthy source!

Think of the potential of such mindlessness. Perhaps it is because of this kind of phenomenon that we had a Nazi Germany, or, on a more personal scale, why we are sometimes too accepting of the influential views of teachers or of books. We owe it to ourselves to develop critical minds and refuse to accept things at face value. (You might like, at this point, to take a look at the section on critical thinking in Chapter 8.)

Enjoy yourselves as you philosophize about the contents of this book or about any other important issues. You have as much right to do so as anyone else.

For starters, here's something for you to sink your teeth into. Many of you reading this book already work in the justice system, upholding the law. Others of you will shortly be seeking employment in the field. What motivates many of you is your respect for law and order. Right? Well, **Martin Luther King, Jr** said there are bad laws and good laws. He considered the law that segregated Blacks in the southern US states to be a bad law because it discriminated between human beings simply on the basis of skin colour. He argued that it is ethical to obey good laws and unethical to obey bad laws—that bad laws should be *disobeyed*. That would mean civil disobedience. It's an interesting viewpoint and one that we should all seriously consider prior to embarking on our careers.

During the preparation of the second edition of the US version of this book one reviewer quite properly suggested that I include a caution here. "This statement," pointed out the reviewer, "appears to make potential criminal justice practitioners believe they will have to make a choice at some time in the future concerning involvement in civil disobedience, or to not enforce a 'bad law.'" In response, first, I am not advocating civil disobedience by peace officers. Second, I am not suggesting, necessarily, that all police or correctional officers will encounter this kind of conundrum. I am simply suggesting that the concept of civil disobedience and the issue of peace officers enforcing bad laws needs to be faced. This is an issue, I would suggest, that you would do well to think through prior to being sworn in.

WHAT IS THE DIFFERENCE BETWEEN MORALS AND ETHICS?

This is a commonly asked question. The term *ethic* comes from the ancient Greek word *ethos*, which means "character." There is really little or no mystery to the difference between morals and ethics. Ethics, as I said earlier, refers to the study of morals and morality, about what's good or bad, right or wrong. It refers to the practice of thinking philosophically about morality. When you read and think about the dilemmas in Chapter 8, you will be engaging in ethics or moral philosophy. Of course, you do not have to do your own thinking on these issues and others like them; you can simply choose to accept the majority opinion and go with that. But even that decision, if you think about it for a moment, is a moral decision. In an important sense, you have made a decision to relinquish your autonomy and, perhaps unthinkingly, buy into the beliefs of the group or society.

Morals, on the other hand, simply refers to human behaviour. Morals have to do with how individuals relate to each other and to their environment. Ethics is the study of those morals. The word *moral* comes from the Latin word *mores*, which refers to social habits or behaviours.

Now that you understand these distinctions, you should be aware that modern philosophers and people in general tend to use these two terms interchangeably. We talk about a person's ethics as we would talk about her morals, and vice versa. So do not be overly concerned about the fine differences between the two.

WHAT IS A MORAL DILEMMA?

In Chapter 9, you are presented with a number of moral dilemmas. Indeed, you will find them lurking in all of the chapters! So it is pretty important that we have some understanding of what a moral dilemma is. We know we all experience them, but what are they? What constitutes a moral dilemma? Are there some characteristics that all moral dilemmas have in common?

A moral dilemma occurs when we are faced with a tricky decision about a complex problem. Generally, the problem is one that does not have a self-evident answer. We become puzzled and worried because we don't know what we should do in the situation. If we decide to do one thing, there is a problem, and if we decide to do something else, there is also a problem. In other words, there may not seem to be any one right answer.

Let's consider the example of whether or not you should tell your best friend about a one-night stand you know her fiancé recently had. The husband-to-be is also a good friend of yours, and *he* knows that *you* know of his indiscretion. He begs you not to say anything, saying that it was a stupid mistake, a spur-of-the-moment thing done out of weakness, and that he loves his partner and wants to marry her and spend the rest of his life with her.

Now, for some of you, this situation may not constitute a dilemma. You may feel strongly that he is a dirty rat and that your friend has a right to the information that he cheated on her. Others of you may feel equally strongly that what happened is none of your business and that two maxims apply: 1) It's always better to let sleeping dogs lie (excuse the pun!); 2) What a person doesn't know can't hurt her (or him).

For still others, however, the solution may not come quite as readily. Many confronted by this situation will agonize, and lose sleep even, over what they should do. The feeling that many have when confronted by such a dilemma is that they are damned if they do and damned if they don't: in essence, that the situation is no-win.

There are at least two lessons here. First, if you and I are faced by the same dilemma, perhaps only one of us will agonize over it. So what strikes you as a dilemma may not strike me in the same way. We should not judge each other, because we clearly come at the problem from two entirely different perspectives and sets of life experiences. For example, if I have been cheated on in the past, I might come more quickly to a decision about what I should do! Also, the right solution for you may not be the right one for me. If that is so, I should neither judge you for your decision, nor attempt to influence you.

For this reason, we've got to be careful at moments like this. When a friend approaches us and confides in us the nature of the dilemma she is facing, we would probably be wise not to be drawn into giving her advice.

The point this book makes, however, is that although we will not necessarily come away with answers we can learn to philosophize *better*, and to think more clearly about dilemmas that confront us. I hope as you read this book you will learn about skills, ideas, and tools that may help you. These tools are discussed in Chapter 8, and if you take a close look at them, you may find one or two particularly useful in working out your own dilemmas.

Something we should be wary of is responding emotionally to a difficult problem. While we cannot entirely discount our emotions, of course, we would be wise to become more skillful in *thinking through* these issues. The tools in Chapter 8 may provide a framework for that thinking.

WHAT IS A WISE PERSON?

Of course, if I thought I was able to offer you a definitive response to this question I would be showing how *unwise* a person I really am! We should wonder about a person making such a claim.

Perhaps the most we can hope for is to identify some qualities a wise person might have and, on the other hand, suggest some things that a wise person is not.

How many times have you heard people waxing eloquent on the significant problems confronting those of us who work in the justice system? There are people who seem ready to offer simplistic solutions to enormously complex problems, who would have us believe they are all-seeing and all-knowing, and who are convinced that the rest of us poor souls are dumb bunnies who wouldn't see the simple solution if it jumped up and bit us on the nose.

I may have it all wrong, but I often tell my students that an expert in the justice field is someone who knows best what *doesn't* work. (And, as you might imagine, many of my colleagues feel this is far too pessimistic a view to take. Do you agree?). Of course, on the contrary, we often hear of different individuals making absolute claims about what *will* work: "We're too soft on offenders"; "Lock 'em away and throw away the key"; "The justice system is too harsh."

Now, I don't want to sound too defeatist here, but at one point or another virtually all of the "new" methods for dealing with offenders have been tried—and found wanting. So anyone making claims that he knows answers to complex problems may be the antithesis of what we consider to be a wise person.

The great philosopher **Socrates** was brought to trial in Athens in 399 BC for supposedly speaking out against the gods of his day and for corrupting young minds. He was put to death, forced to drink a cup of poison called hemlock. His unpopularity had its roots primarily in his ability to expose those who claimed to be wise but were not. In his defence, he said something quite significant. He had been talking about a politician whom people had thought wise but who had been exposed by Socrates to be a fraud. He said,

> Well I am certainly wiser than this man. It is only too likely that neither of us has any knowledge to boast of; but he thinks he knows something which he does not know, whereas I am quite conscious of my ignorance. At any rate it seems that I am wiser than he is to this small extent, that I do not think that I know what I do not know ("The Apology," in *The Last Days of Socrates*, p. 50).

A wise person, then, is probably someone who would seldom if ever claim to "know" much about anything, particularly in an absolute sense. He would be a person who would be cautious about making any kind of absolute claim. He would be thoughtful and well-read about any topic before venturing an opinion. He would be open-minded and not afraid to look at evidence that contradicts or competes with his current views. He would have a critical mind and always be on his guard against unthinking acceptance of ideas not carefully examined. He would certainly resist any pressure or temptation to hold what we sometimes call a "blind belief."

It would be a comfort, wouldn't it, to know that our elected officials, our law makers, act wisely when enacting new laws or repealing old laws. Can you think of any instances of laws becoming enacted or repealed based on what you consider to be faulty reasoning or a demonstrated lack of wisdom? Can you think of any new laws that have been passed based on an emotional appeal to a vocal segment of our society?

As you respond, be careful not to bring into the equation your own emotional baggage. Be moderate and thoughtful in your response.

Take care, as you wrestle with the dilemmas presented in this book, that you don't come on too strong about the "right" or "wrong" response. Try as hard as you can (and I know from personal experience that it is terribly difficult) to be humble, and to be conditional and measured in your thinking.

Of course, to seek to become a wise person, as an end in itself, is no doubt doomed from the outset. We can certainly attempt to conduct ourselves in such a way that we aspire to some of the qualities of a wise person. But to seek to become a wise person as an end in itself will almost certainly result in frustration and disappointment. Wisdom may be a bit like greatness in this regard. Generally, truly great people do not seek to become great; they have greatness thrust upon them. It is as though greatness chooses them, not the other way round. And so it might be with wisdom. Seek it for its own sake, and it almost certainly will be elusive. Seek to be honest, open-minded, cautious about the truth, and humble as you search for knowledge and meaning, and wisdom may find you.

Reflections

When is learning not learning?

What do you think of the idea that when we are thinking deeply about important issues, we are philosophizing?

What is your definition of wisdom? Of a wise person?

What do you think of what Socrates said in his defence?

Most of us, at times—and I very much include myself here—have come on strong about having "answers" to complex problems. Why do you think it is that we set ourselves up in this way?

What drives our need to set ourselves up as experts?

There is an expression I have heard, and perhaps you have heard it too, that goes something like this: "The more we know, the more we don't know." What might this mean?

IDEAS IN A CONTEXT

Perhaps the best way to examine these ideas, as I have already hinted, is to look at them *in a context*. That is, we take an issue that is relevant to the justice field and examine it from the perspective of some of the different ideas and theories that have been written. Of course, this examination, as I suggested in Chapter 1, will be a cursory one—it will be neither in-depth nor exhaustive. We will touch on only some of the ideas out there.

Keep in mind that you may feel the context is introduced in a provocative way. The approach is deliberate, but it should in no way be construed as my peddling any particular point of view.

The Context

Here's the context. Every society has its criminal element. It is a practice in most societies, particularly ours, to deprive offenders of their freedom when they do something wrong (some of us may think not for long enough). The purpose of locking them away, we say, is to give them time for reflection, to provide a safeguard for the rest of us in society while they reflect, and to punish them for their wrongdoing. Underlying our system of justice is the idea that people are free moral agents, and that if they knowingly commit an unlawful act they will be punished accordingly.

So let's examine the practice of locking aberrant members of our society away, and how this practice squares with some of the philosophical theories. Keep in mind that we will just touch on these ideas as a way of getting the thinking juices going, as it were. None of the arguments or ideas are meant to be definitive.

Libertarianism

Libertarianism makes the assumption that we all have free will to make our personal choices, and that based on the choices we make we should be praised or blamed. If we do good things, we should be praised, and if we do bad things, we should shoulder responsibility for those acts and take the consequences. William James (1842–1910), who was both a philosopher and a psychologist, was one of the main exponents of this point of view.

If we think about it for a moment, our system of law is predicated on this basic assumption that we are what philosophers sometimes call "free moral agents." This means that we should consider our options carefully before we act, because we will be held accountable. When we do something wrong and are charged with an offence, the court will establish whether we are to blame, and much of the court's decision will rest on whether it can be shown that we acted with criminal intent. The notions of *mens rea* and *actus reus* loom large in criminal law. *Mens rea* refers to the idea that we acted purposefully and with evil intention. *Actus reus* refers to that which is forbidden by law.

Put simply, the theory goes something like this:

Fred broke the law.

Fred knew that he was breaking the law.

Fred could have acted otherwise.

Because of the choice he made, Fred deserves blame and punishment.

It is just and fair to punish Fred because he freely chose to do a bad act.

If this kind of thinking is correct, and it may well be, then our system of law would appear to be just and fair. The maxim "Let the punishment fit the crime" appears to be a reasonable one if we are, indeed, free to make individual choices. Potential offenders should know that society will not respond kindly when they violate the freedoms of the rest of us.

Certainly, it is only by thinking in this way—that is, that we are free moral agents—that we can make much sense of our lives, our accomplishments, our system of rewards and punishments, praise and blame, accountability and personal responsibility.

What do you think? Do you think we are free to make our choices and that we should be held accountable for those choices?

Reflections

What do we mean when we say a person is responsible for her own actions?

There is, however, another way of looking at this issue of whether we do, indeed, have free will—the school of thought called determinism.

Determinism

Rather surprisingly, perhaps, novelist Mark Twain of *Huckleberry Finn* fame was a leading advocate of **determinism**. Here's something he wrote:

> Man is not to blame for what he is. He didn't make himself. He has no control over himself. All the control is vested in his temperament—which he did not create—and in the circumstances which hedge him round from the cradle to the grave and which he did not devise.... He is purely a piece of automated mechanism as is a watch, and can no more dictate or influence his actions than can the watch. He is a subject for pity not blame... (Quoted in Grassian, Victor, "Moral Reasoning," p. 166.)

For Twain, then, we are victims of our circumstances. We are shaped by heredity and by the particular and peculiar environment we experience. We have no control over the process. To all intents and purposes, human beings are like dried leaves blown here and there by a winter wind.

We are subject essentially to the continuum of cause and effect. Each one of our behaviours is conditioned or caused by something else. The bottom line is that, as human beings, we can become no other than what we become. Sometimes we refer to this kind of thinking as "fatalism"—that what will be will be and there's not much we can do about it. Also, those of you interested in theology will no doubt be familiar with the term *predestination*. This concept, derived principally from the writings of the theologian John Calvin, speaks to the idea that God has determined for each one of us ahead of time what our fate will be in this life and in any life following this one.

Now, if the determinists are correct, there may be far-reaching implications for our justice system. For example, we may well ask ourselves whether it is fair and just to punish someone for committing a murder or rape when, in fact, he could have acted no differently. The thinking is that the act, in a sense, chooses the person, not the other way round. The person becomes who he is, including all of his choices, because of his unique birth and life experiences.

Let's have a go at applying this way of thinking to the justice system. Have you ever considered how our courts sometimes work? Would you agree that there might be some substance to the idea that an accused person dressed very well and with a reputable lawyer at his side may experience a more favourable verdict than a person who is unkempt, has few if any social graces, and is poorly represented? Now I know that I am stepping into shark-infested territory here, but I am doing so, as I said earlier, to get the intellectual juices going. I am being deliberately provocative, so it is important you understand that this discussion is not to be construed as an attempt to influence you.

Let me tell you about an experience I had many years ago when I was working as a parole officer. I was on court duty. Often—and I have to be careful here not to overgeneralize or to become guilty of gross stereotyping—a criminal court is populated by what we call the underclass in our society. They are the "have-nots": the poor, the downtrodden, the institutionalized—the marginalized members of our society. From time to time, however, sitting in the courtroom awaiting their turn before the judge are what appear to be more privileged members of our society: professional folk, people who are well-groomed and attractive.

On that particular day, I looked around the courtroom and noticed two attractive and immaculately turned-out individuals, a man and a woman—husband and wife. It turned out that they were both employees of a major airline, and he was before the court on a charge of shoplifting. The charge was that he was in one of the local grocery stores and stole $170 worth of groceries. Apparently, he wore his airline overcoat and stuffed the numerous inside pockets with pork chops, butter, candies—and the list went on. When his case came up, his lawyer told the court what a law-abiding, productive member of society his client was, but that due to tremendous job pressures he had found himself acting out of character on the day in question. He didn't know why he had done what he did, he was deeply remorseful, and his professional and lovely wife was in court that day to show her support. The prosecutor responded in a conciliatory manner: the accused was given a conditional discharge.

By contrast, the next person to appear before the judge was a scruffy, unkempt individual with a previous criminal record. He had declined representation. He said, yes, he had been in the grocery store (the same one as the airline person). He was hungry, so he took a loaf of bread. He told the judge that he was pleading guilty and that he should be sentenced so that he could get it over with. The judge decided that this request amounted to insubordination and disrespect to the court and ordered the accused remanded in custody pending the preparation of a pre-sentence report. This individual spent the following six weeks in custody.

Let's think of this case from the perspective of the determinist. In one sense, logically, criminal courts ought to look more favourably rather than less favourably on society's underclass. It might be argued that these individuals know no better. Their conditioning, often, has been to survive by taking from the "haves" in society. And, often, they feel little or no remorse about doing so—it is the way they have learned to survive.

On the other hand, the airline crew member presumably had a much better life experience, with parents who loved him, a wife who cared for him, an employer who valued his services. His life experience was probably one of good schooling, regular church attendance, and community leadership.

Logically, it could be argued that the courts should come down much harder on those of us who ought to know better. After all, good fortune has smiled on us in terms of our genetic make-up, and the experience of enjoying a supportive, constructive, nurturing environment coupled with the benefit of good modelling in terms of respect for the law and the rights of others. The courts, however, may work the other way, as they did the day I was observing.

Let me give you another example, only this time, rather than looking at the courts, we will consider how police officers might sometimes be inclined to do their jobs. This is confession time! Two days ago my wife and I were in a hurry, and I was driving over the speed limit. I saw the cruiser, but too late—my goose was cooked. The lights went on; he followed me and pulled me over. Was I aware that I was driving at 127 kph in a 90 zone? *Yes.* Was there a reason for my speeding? *Not really, but we were running late.* The officer asked to see my driver's licence and insurance. He disappeared into the cruiser, reappeared five minutes later, and told me that I could have been facing a fine of $227, but that he was going to give me a break. The fine would only be $47 with no loss of points.

Notwithstanding my gratitude for the "break," I have to wonder what the response would have been had my vehicle been old and I been scruffy? Would I still have received the benefit of the break? You judge. Interestingly, about half an hour later as we continued our journey, we saw the lights of two cruisers ahead of us. A truck had been pulled over, and the police officers were questioning the driver on the side of the highway. He did not present the most attractive of images. "Oh, oh," said my wife, "that poor b— is going to get the book thrown at him." Now maybe he got a break, too, but I rather doubt it.

Think about your life circumstances. To what extent do you think you are primarily responsible for them?

Reflections

Think about what you had for breakfast this morning. Think about the clothes you are wearing today. What factors, if any, played a part in the "choices" you made?

In light of the foregoing discussion, is our justice system a fair one? If you think it is, why? If not, why not?

What if the purpose of our having laws was to prevent the "have-nots" from taking from the "haves"? If laws were there essentially to protect the privileged in our society, what then?

Is it fair to hold as morally responsible for their actions individuals who don't know what moral responsibility is?

What do you think my response to the "break" should have been? Perhaps a more honourable response might have been to say, "Officer, you caught me fair and square. I do not deserve any break. Give me the ticket, and I'll send off the cheque for $227."

Utilitarianism

Jeremy Bentham (1748–1832) was an English philosopher and John Stuart Mill (1806–1873) was one of his students. Bentham's primary interest was in legal reform. One of his best known works is his *Introduction to the Principles of Morals and Legislation.* Mill himself was a philosopher as well as an economist and he was primarily responsible for coming up with the term **utilitarianism** to denote the moral theory developed by his teacher. (The term Bentham had used was *the principle of utility.*)

Reference is made to this principle in Chapter 8, Tools for Moral Decision Making.

The principle of utilitarianism, formulated by Mill and Bentham, is also known as the principle of benefit maximization. Stated simply, the principle contends that a decision made or an action taken by an individual or a state is justified or not justified by its consequences. If a decision or action results in less pain, or greater good, or the world generally becomes a better place, then the decision or action is justified. It is the consequences that matter rather than the inherent rightness or wrongness of a particular decision. So here it could be argued that if my telling a lie results in the world becoming a better place, the lie is justified.

Let's examine this principle with respect to the jailing of lawbreakers. It could be argued that if our society, generally, is a less dangerous place because of the existence of correctional institutions, that is a sufficient reason for having institutions.

But what if it could be argued that our society is a less dangerous place—but only *temporarily*? What if it could be argued that society becomes a *more* dangerous place in the long run; that in institutions offenders learn new tricks, and more often than not their anger and resentment toward society build because of their experience of incarceration; that when they get out, we're all fair game for their next bitter, anti-social act?

There are some serious implications here for *why* and *how* we incarcerate. If our methods of incarceration are designed to be repressive and harsh, might they become counterproductive in the long term because most offenders at some point will be released? We may do well to ask what kind of people are going to be let back into our communities.

Reflections

Would you agree that generally the consequences of treating people reasonably are better than those of treating them unreasonably? Think about your own experiences and how you respond. What might be the implications here for policing and corrections?

What is your view on this issue? Do you think the principle of benefit maximization should be considered a guide for the kind of justice system we want?

If a repressive justice system can be shown ultimately to lead to a less safe society, would that be sufficient reason to become less vengeful in the way we deal with offenders? Or not?

What if, in your view, the justice system is too soft on offenders and there is no real deterrence anymore because the courts are responding too leniently? How would you square this view with the idea that a repressive, punitive system may cut society's nose off to spite its face?

The Principle of Equal Respect

This principle is purported to be central to our system of justice. Put simply, it exhorts us to treat each other with equal respect, requires that we all be considered equal before the law, and says that purely and simply on account of our being human, we have intrinsic worth. It follows then, that *all* human beings, irrespective of education, class, colour, or any other differences, are of equal worth.

The utilitarians argue that we should treat each other well because the consequences of doing so would probably be better. The principle of equal respect, on the other hand, would require us to treat each other well simply because that is the reasonable and decent thing to do. Individually, it is our being human that buys us a place, as it were, in the equal-respect club.

In their book *The Ethics of School Administration*, Kenneth Strike, Emil Haller, and Jonas Soltis make the point that this principle comprises three basic ideas:

- That we treat people as *ends in themselves rather than means*. This means that we are morally obligated to consider the welfare of all people.
- That we accept all humans as *free and rational moral agents*.
- That no matter how people differ, as moral agents they are of *equal value* and are owed all basic human rights (p. 17).

If we think about this principle for a moment, we can see where the legal notion of due process has its roots.

Now, what might be the implications of the principle of equal respect for the justice system? Well, let's think for a moment about arrest procedures. Perhaps it would be appropriate, as and when time permits, of course, for officers making an arrest to think about how *they* would feel if they were the accused. The thinking process does not necessarily have to be a long and laborious one; it only takes an instant. What if this person being arrested were my father or mother, sister or brother? What would I wish their arrest experience to be like? Of course, our response would obviously be dictated at least to some degree by the behaviour of the person being arrested. But nonetheless, if my dad were being uncooperative when being arrested, wouldn't I still expect him to be afforded certain courtesies?

What if I am a correctional officer and I am supervising an inmate who has just received a Dear John letter. His wife has informed him that she has moved in with his best friend, also an offender, who himself was just released from jail last month. When the inmate reads the letter, he becomes very angry and aggressive. He is banging around in his cell and letting go with all the expletives he can muster. You ask him to quiet down, and he tells you to go away (in a biological manner of speaking!). The rules require you to book him. What are you going to do? Are you going to respond legalistically? Or are you going to take a moment to find out what's happening, and then, after you have done so, are you going to attempt to put yourself in his skin? How would you feel? Are you going to respond to him as you would hope other officers would respond to you in the same situation? Or is this not what the principle of equal respect is all about?

A key question we might ask is this: Are we all really equal? Or might there be one law for some, and another law for the rest of us?

Of course, there's much more we could say about the principle of equal respect and its implications for the justice system. What is important now is that you do your own thinking about those implications.

Reflections

If we follow this principle to its logical conclusion, might it mean that we should respond to everyone—whether in the courts, out on the street, or in our institutions—in a similar way irrespective of their circumstances, including age, race, gender, or sexual preference?

If we think there is a place for discretion, for example, in the criminal justice system, how is it possible to square that with the principle of equal respect?

The codes of conduct of many law enforcement agencies talk about applying the law without fear or favour. Sounds like the principle of equal respect, doesn't it? Where do you stand on this?

How do you think you respond to society's underclass? They are the people with whom you work primarily—the poor, the disenfranchised, the minorities. Do you think of each one of them equally because they share your humanity?

PHILOSOPHICAL THOUGHT

As I said in the Introduction, this is where we are going to look briefly at some of the big names in philosophical thought. Of course, what follows are brief references indeed, but I hope they will be enough to whet your appetite and fire a curiosity in you to find out more.

Plato

Plato lived from 428 BC to 348 BC, so he lived to be quite an old man. He is best known for a famous work called *The Republic* (sometimes known as Plato's Republic). As a young man, Plato established a friendship with Socrates, who is the next person we are going to talk about. Socrates was both Plato's mentor and his teacher.

The Republic, briefly, is an account of Plato's thoughts on society and how society would best be shaped. It is a dialogue in which arguments and points of view are put forward and discussed. Plato is particularly interested in the place of justice in society. And one of the main injustices, as he sees it, is the gap and the conflict between the rich and the poor.

Francis Cornford, a philosopher, commentator, and writer on Plato and *The Republic*, says that the main question to be answered in *The Republic* is this: What does justice mean, and how can it be realized in human society? The Greek word for *just*, says Cornford, can have many meanings: observant of custom or of duty, righteous, fair, honest, legally right, lawful...what one ought to do. Justice, says Cornford, has to do with the individual's conduct as it affects others. The relevance of this kind of discussion to those preparing for a career in the justice system is obvious. As you wrestle with the dilemmas presented later in the book, it is important to include the concept of justice in your thinking.

Plato examines what would constitute a just state and how members of that state would behave in terms of their relationships with one another. What is good is discovered through the use of *reason*. The thinking and knowledgeable individual is the one who will be a good person and do good acts. Bad or immoral acts, on the other hand, are the result of ignorance.

Here is one of Plato's thoughts—among many, many others—that might have relevance for us now and for how we do our jobs as peace officers: "Our object [or objective] in the construction of the state is the greatest happiness of the whole, and not that of any one class."

Perhaps Plato's biggest contribution is that he set an example for us. We should think carefully about the kind of society we want to live in, the kind of people we are, the kinds of behaviours we engage in and how these behaviours affect others.

Socrates

Socrates was a philosopher who lived from 469 BC to 399 BC. He was a funny-looking fellow, had a much talked-about marriage and several children, and for all his life was as poor as a church mouse. Eventually, he was put to death by being forced to drink a cup of poison. The charges against him were that he spoke out against the gods that the citizens of Athens believed in, and that he corrupted the minds of the young. Plato's mind would have been one of those that Socrates was accused of corrupting. When Socrates died, Plato said, "This is the end of our comrade, a man, as we would say, of all then living we had ever met, the noblest and the wisest and the most just."

Socrates never wrote a word himself; his ideas and his wisdom were captured and written about by Plato.

Even though he died well over 2 000 years ago, his thinking is still influential today. He is famous for developing what we now call the Socratic method of teaching, something you may experience, in fact, with your teacher or instructor. It was certainly an experience that Plato would have enjoyed when he was a young student of Socrates's. The Socratic method simply refers to the free and open discussion of ideas, and this, Socrates felt, was the only way to achieve knowledge and enlightenment. In Socrates's day, these discussions would more often than not take place on street corners and, indeed, anywhere there was a gathering of folk interested in a dialogue about important matters.

One of Socrates's legacies is one simple but profound thought: "The unexamined life is not worth living." He is saying, essentially, that if we elect to be dead from the neck up, if we choose not to use our reason or intellect, our lives are going to be less meaningful. It was important for Socrates, and I would suggest it is equally important for us, that we face squarely issues concerning human existence. It will take courage, but at least we will feel alive and more complete. And, as a result, we will probably become more knowledgeable.

Becoming more knowledgeable was important for Socrates because he believed strongly that there was a close link between *knowing* the right thing to do and *doing* the right thing, and that thinking about how we should behave leads to better behaviour. He also believed there is a close link between behaving morally and being happy. Perhaps paradoxically, however, he felt strongly that the wisest person is the one who realizes how ignorant he is.

One final thought about Socrates. Both he and his friends were keenly aware that Socrates had been imprisoned on trumped-up charges and that his imprisonment and death sentence were unjust. His friends, particularly one close friend Crito, wanted to help Socrates escape. Socrates would not hear of the plan, because to escape would be to go against the laws of the city. He said to Crito, "Do you imagine that a city can continue to exist and not be destroyed if the legal judgments in it have no force, but can be nullified and destroyed by individual persons?" Clearly, Socrates was against anarchy. His respect for the law was absolute.

Earlier we noted what Martin Luther King, Jr had to say about bad laws and good laws. He said that we should respect and obey good laws, but that we have a moral obligation to disobey bad or unjust laws. Where do you stand on this issue? Are you with Socrates or with Martin Luther King, Jr?

Aristotle

Aristotle was one of Plato's students. He lived from 384 BC to 322 BC. He was an argumentative student who had many disagreements with his teacher. (Not that there is anything bad about this, of course. One of the greatest compliments a student can pay his teacher is to feel comfortable enough to engage in open, honest discussion with him.) He was both a scientist and a philosopher. His science background led him to be quite practical and down-to-earth in his thinking, and because of this background, he devised what we call "formal logic." You may also hear it referred to as Aristotelian logic. Logic simply refers to rules of argument or of applying correct reasoning and making valid inferences.

Here is one of the more common examples of formal logic you may have seen in other books:

Socrates is a man.

All men are mortal.

Therefore, Socrates is mortal.

We also call this kind of argument a *syllogism*. You can see that the conclusion follows from the premises (the first two statements). If the premises are true, as they are here, then the conclusion is also true. But if the premises are false, then the conclusion is also false, even though the argument has a logical form. So in arguing correctly, there are two important considerations we should keep in mind: the truth of the premises and the form of the argument.

If you remember what we said about the determinists' point of view, here is how they would frame their argument syllogistically:

Every event has a cause.

Every human choice or action is an event.

Therefore, every human choice or action is caused.

Here, the argument is a sound one with respect to form. However, we do not know for sure if the premises are true, so we cannot be sure if the conclusion is true.

I remember seeing a film on moral philosophy once—I can't remember its name, unfortunately, so I can't properly attribute it here—in which the narrator was talking about good and bad argument, sound and unsound reasoning. He used the following example to show sloppy and unacceptable thinking: If I think abortion is wrong because killing is wrong, then there would be something wrong if I subsequently declared I was in favour of war. There is a basic flaw or inconsistency in my argument because the major reason I cite in support of my first viewpoint does not carry over to the second.

Think about your own reasoning as you go through this book! Are there any inconsistencies in your arguments? Are your thoughts logically connected or are they contradictory?

This brief overview of Aristotle's contribution is not the place for an in-depth discussion of formal logic. This is just a minute taste. But you might like to take a course in logic sometime. You may find that it will help to clarify your thinking.

As well as this significant contribution of logic, it is important to note that Aristotle was what we call a *synoptic* philosopher. He was a man who had a broad range of interests. He was interested in all the sciences and, of course, he was passionate about philosophy. He attempted to integrate all of his interests into his thinking and writing. By so doing, he was attempting to develop what we call a "world view" by integrating empirically known information with rational thought about the world.

Plato, Socrates, and Aristotle are what we call ancient Greek philosophers. We will now move forward in time to look at one of the "biggies" from the seventeenth century.

Thomas Hobbes

Hobbes was an English philosopher who lived from 1588 to 1679. I have chosen to include him because his idea about something he called a *social contract* has relevance to how we think about our systems of government and justice. His main work was called *Leviathan*, and in it he developed a system of thought called *ethical egoism*. Ethical egoism, as you might guess, refers to the notion that people ought to act in their own interests.

It is Hobbes's thinking on the social contract, however, that is most important for us, because the social contract refers to the rules of justice, how they came about, and how we should respond to them. The rules of justice developed because people agreed that they were necessary for a stable society, that they should be obeyed, and that a government that enforces them should be obeyed and respected, too. What it amounts to is that we have all agreed to be subject to the constriction of these rules because it is in our best interests to do so. We give up some freedoms in order to gain a greater measure of freedom. If I obey the rules, some of which will necessarily restrict my freedom, and you do too, we can all live happily ever after, right? If you decide not to obey the rules, then it appears only reasonable that you should be held personally accountable and made to obey.

This contract, for Hobbes, is a kind of glue that will hold a society together. There will be relative order and peace, and we will be safe from each other because we have all bought into the plan. Without the contract, we would be at each other's throats and, essentially, we would be living the sort of brutish existence to which the schoolboys sank in the book (and subsequent film) *Lord of the Flies*.

The social contract, then, is essentially self-serving. It protects you, but, more importantly, it protects ME.

Let's consider this theory for a moment. If Hobbes is correct in his thinking—and he was generally supported later by John Locke (1632–1704) and Jean-Jacques Rousseau (1712–1778)—the theory makes sense if we all have "stuff" to start with. But what if I don't have anything worth calling my own? Why should I agree or be subject to a social contract? Such a contract would be in the best interests of those who have something worth hanging on to.

Who knows where the truth lies? But it is a conundrum to which we should give some serious thought as people charged with the responsibility of upholding law and order. Keep in mind that mine is a simplistic and rudimentary explanation of some of Hobbes's thinking. Let me encourage you to do more exploring on your own.

Immanuel Kant

Immanuel Kant was a German philosopher who lived from 1724 to 1804. His work has contributed greatly to philosophical thought, but you should be warned that if you decide to read him for yourself, it can be a distinct challenge. All of his works, of course, have been translated from German to English. His most significant work is *Foundations of the Metaphysics of Morals*.

I have decided to include Kant in these very brief synopses because of the importance of one of the concepts he developed in the *Foundations*. I refer to his idea of the categorical imperative, otherwise referred to as the theory of moral obligation, an idea that has great significance for justice practitioners and anyone who works for the public good. For him this idea constitutes an overarching, all-encompassing moral principle.

Kant was very much opposed to the notion of *determinism* which we discussed earlier in the chapter. He asserted his belief in free will, because otherwise life, in general, and a sense of moral obligation, in particular, would not make sense. We can only talk about individuals having a moral obligation to do something if in fact they *can* do so. What separates humans from animals for Kant is the human capacity to be rational, which, in turn, gives us the ability to make moral choices.

So what does Kant mean by the *categorical imperative*? Of course, he expresses many ideas about this notion in his writings, but one of his ideas is that morality requires us to treat all human beings as ends in themselves and never as means to an end. Clearly, there is a link here with what we call the *golden rule* of Christian theology, which is: "Do unto others what you would have them do unto you."

Many of you will agree that what inspired you to become a police officer or a correctional officer was a desire to make a difference, and to conduct yourselves with professionalism, courtesy, and respect when dealing with the less desirable elements of our society. And you will know that when the situation demands it, you will be prepared to put your life on the line to help others. Isn't it this that drove firefighters and police officers to help in any way they could on September 11, 2001? Is it this kind of selfless act that the categorical imperative is all about? It's about having a sense of duty, of obligation, such that you cannot help but do what you feel you must do.

One of the reviewers for this second edition of the book, when suggesting I include this segment on Kant, talked about a justice student graduate she had who recently rushed into a blazing building and saved two lives at great risk of losing his. She said he did not do it for personal glory, or even because it was his job. It was his duty to the people he was there to protect and "he believed that into his very soul."

So Kant sees morality inextricably linked with the idea of duty, and this idea will only make sense if we are free to do what it is we consider our duty. You might ask yourself the question whether your desire to become a peace officer has anything to do with your having a sense of duty.

Finally, we will take a look at a contemporary philosopher.

Jean-Paul Sartre

Jean-Paul Sartre was a French philosopher who lived from 1905 to 1980. He was what we call an *existentialist*. Existentialist philosophers write about the human experience, principally about human freedom, and about the importance of our being responsible for our actions.

Sartre leaves no doubt about his view that we are and must be responsible for our actions when he says,

> Existentialism's first move is to make every man aware of what he is and to make the full responsibility of his existence rest on him (*The Big Questions*, p. 160).

The view that we do not have free will, for him, is an outrageously false one. It is a cop-out. You may remember our discussion earlier in the chapter about determinism. For Sartre, determinism is the ultimate cop-out. I must carry full and complete responsibility for my choices because nothing dictates them except me. But, for him, there is a downside: with freedom comes awesome responsibility, and consequences!

Sartre talks, in fact, of our being "condemned to be free." It is almost as though if we could avoid the responsibility of freedom, we *should*. But we can't. Our choices and their consequences affect not only us and those close to us, but also, conceivably, the entire world. This is a gobsmacking thought. It is a thought that ought to haunt us whenever we make any decisions of substance.

For Sartre, there is no escaping this responsibility. As James Christian, in describing Sartre's views, says in *The Art of Wondering*, "There is nothing to help us—because the moment we become conscious of what we are, then we become responsible for everything we are and do" (p. 241).

It is interesting to note that Sartre's views on freedom may have arisen from his experience during the Nazi occupation of France in the Second World War. He wrote the following:

> We were never more free than during the German occupation. We had lost all our rights, beginning with the right to talk. Every day we were insulted to our faces and had to take it in silence.... And because of all this we were free. Because the Nazi venom seeped into our thoughts, every accurate thought was a conquest. Because an all-powerful police tried to force us to hold our tongues, every word took on the value of a declaration of principles. Because we were hunted down, every one of our gestures had the weight of a solemn commitment.

Of course, as spirited as Sartre's defence is of the idea that we have freedom of choice, that does not mean to say he is correct. As we have noted earlier, there may be significant difficulties with this point of view.

SUMMARY

We have not even scratched the surface of philosophical thought in this chapter. But your hand has touched the surface, and it is up to you to explore further.

Of course, there are many other giants in the field of philosophy who have contributed much to our understanding of who we are as human beings, and of the world in which we live. I hope you will be encouraged to continue reading on your own. But beware: some of these writers and thinkers can be terribly difficult to read. For many of them, you will need your headache pills handy. I remember one friend saying to me that many philosophers operate on this principle: Why say it simple when you can say it complicated!

In any event, I hope this discussion has stimulated your thinking. When some of the greatest minds cannot agree on important issues, it is little wonder that we disagree with each other in our conversations. That is the way it should be. In a healthy debate there should always be room for disagreement and difference of opinion. That is how we learn from each other.

I hope such learning will be your experience as you continue to discuss the ideas in the rest of this book. Do not hesitate to refer back to these ideas if you feel they can shed some light on a problem you are thinking about or discussing. Remember, good learning is about taking insights you gained in one context and applying them in another.

chapter three

The Ethics of Teaching Ethics in Justice Programs

In taking the ethics course, I have now established a better understanding of who I am, what I stand for, and for what reasons.

—Journal entry. Justice student in ethics course.

I have a hard time separating ideas and feelings.

—Journal entry. Justice student in ethics course.

Feelings from time to time are going to be hurt. We don't want to hear or face the possibility that our opinions may be biased or prejudiced.

—Journal entry. Justice student in ethics course.

One thing that I didn't like about the course [ethics] was that the teacher never really shared his views with us. I'm not sure it is fair to expect the class to share their personal feelings with an instructor who will not engage in any sort of self-disclosure. I would have liked the instructor to allow the class discussions and then maybe take five minutes or so at the end of the class to share his thoughts with us. Kind of like the Final Thought on *The Jerry Springer Show* (ha, ha!).

—Journal entry. Justice student in ethics course.

CHAPTER OBJECTIVES

After reading this chapter, you should be able to

- Explain what we mean by **teaching ethics**
- Explain what the purpose of ethics education might be
- Identify reasons for including ethics education in curriculae
- Explain what is meant by **advice-giving** by a teacher
- Identify the situations in which advice-giving might be appropriate
- Explain the collaborative approach to ethics education

THE RATIONALE FOR CHAPTERS 3 AND 4

Chapters 3 and 4 will be more relevant to those groups using *Reputable Conduct* as part of a classroom experience. The lone reader may find the chapters interesting also, but they clearly would not impact her in quite the same way.

It is quite unusual to include this kind of discussion in a text book. Some readers, especially teachers, have suggested that these chapters would be a good resource for the teacher but that they may have little relevance for the student. For example, a reader made this comment: "This chapter seems to me to be more of a tutorial for the instructor than the student...a kind of instructor's manual."

My view is that chapters 3 and 4 could, indeed, be construed as a tutorial or a manual, but they are intended to be that for *both* teacher and student. The discussions in these chapters, if you are using this book as part of a classroom experience, are there to enhance the process of learning that both teacher and student will be undertaking *together*. The process of "teaching" ethics can be a veritable minefield of potential difficulties, and it is better for you to have a discussion early on about what mines may be laying in wait, rather than stepping on one part way through the process and having the experience spoiled for everyone concerned.

One of my students wrote the following: "People do not realize how narrow-minded they are until they are hit in the face with their narrow-mindedness." Now here she was not only referring to fellow students but to herself also. As you can imagine, suddenly seeing oneself as "narrow-minded" is probably a painful experience for most of us. That is why it is critically important to understand the *process* of ethics education before you embark on it. In this way, you will be less likely to react personally to the class discussions and will be better able to place them in a context of intellectual exploration of ideas rather than that of personal attack or exposure.

At this point I am going to take a big risk and explain the primary reason for including these two chapters in the book: Several years ago I had one of my most difficult and painful experiences as an ethics educator. Within three or four weeks of the course start-up I became aware of some very discomfiting dynamics in the classroom. It was clear that some of the students felt that I was coming on as though I knew what was best and that I was discounting of some of their expressed views that they thought were contrary to my own. Here's what one student wrote about what was happening:

By challenging people's thoughts and beliefs and presenting them with questions to the answers they thought they had, John required people to look inside themselves. For some this was a foreign and scary experience, and they were unable to justify their opinions.

In other words, some of the students felt under personal attack when their opinions were confronted intellectually. What I had failed to impress upon them, even though I had expressed the sentiment many times, was that if an idea was challenged by me in class (and those ideas included my own) that this was simply a process of exploration and did not imply my own belief or opinion on the topic under discussion. In other words, it is very, very important for you, both teacher and students, to make a concerted effort to more fully understand what the process of ethics education is prior to embarking on it.

Interestingly, the student who wrote the above sentiment, went on to say this:

When he asked questions that may have challenged my opinions and called for justification, I found it a rewarding experience to be able to rise to the occasion. When I was not able to justify my beliefs I was inspired to seek new information and knowledge, thus learning more about myself in the process.

I guess you win some and you lose some! Seriously, though, it should be the objective of all concerned, including the teacher, to attempt to respond to robust intellectual enquiry in an open and honest way.

Let's consider what another student had to say:

It appears that every time I had to do some soul searching on issues discussed in class were the times I was suspicious of John's objectivity. Doubting myself and some of my beliefs led me to direct doubt toward John. Who better to blame than the man teaching me how to think critically, which must always consist of self-analyzing?... If only I had known at the beginning of this course the self-evaluation I would be performing on myself.

That's why it is so very important for you to discuss the teacher's role at the beginning of the course; likewise it is equally important to discuss students' understanding of that role, and, equally importantly, *their* role.

You have read what some of the students had to say in their course journals about this difficult class experience. Well, I wrote a journal, too. This is the risky bit, because I am going to open myself up and let you know some of the things I wrote. I'm going to discuss my comments and the comments of my students, and I hope this discussion will prepare you for a rewarding experience in your ethics class. I wrote:

Two students visited me this morning to say there were some students in the [ethics] class who were feeling intimidated by me. Students were, they claimed, afraid to speak their minds on ethical issues experienced on their field practicum because they thought they might be judged by me in some way. They also claimed that some students felt that I exerted too much influence on certain of the issues.... I don't quite know how to shake this "influence thing." It has happened in other years so it has to be real - at the very least the perception of it. Apparently, students respect me and, as a result, see me as intimidating and this persists despite a conscious effort on my part to avoid this perception. For example, a) I have commended students for speaking out even though they may be arguing for what they would think they would see me regarding as the "unethical" view, like physically assaulting a prisoner; b) I have repeatedly explained that when I ask more questions of one side of an issue it is simply an attempt to balance the equation—that I have a responsibility to get the class to explore all the sides of an issue; this, the students, are informed, is in no way an indicator of the way I feel about that issue; and, c) students who have argued well for a viewpoint different from mine have been rewarded academically; conversely, students who have done a poor job of arguing for a position I happen to hold have not been rewarded academically, and I have assured them of this repeatedly. [Note: generally, I do not express an opinion unless very much pressed to do so, so the students would only have my word for this.]

As a result of the concerns expressed by the class, I decided to have a discussion with them about the problem. During that discussion, one student raised an issue that had troubled her. She had said in class that "she had problems with Indians," and I had taken issue with her over the comment. She was correct, I had stopped and addressed the comment. Here's what I wrote in my journal:

> In the class I suggested to her that she think hard about her comments because they bordered on racism. She again said that she had difficulty finding much good to say about Indians, and I again suggested she do some serious reflecting on her view. She accused me of judging her. Did I? Is it not my place as a teacher to confront, albeit delicately, socially unacceptable and/or racist views? Is it possible to challenge in a situation like this without judging?

> This class discussion made for a very difficult two hours. It's hard to sort out genuine, reflective critiques from those that are motivated by pique and low grades.

> At the end of the discussion, I came very close to losing my composure, when several students offered heartfelt gestures of support and expressed gratitude for the mind-broadening experience. The applause at the end was somewhat comforting but not completely—for it was difficult to tell if it was inspired by their feeling empathy for me or if they were attempting to atone for their lack of charity.

There, I said I was going to take a big risk! This experience happened several years ago and I have since taken great pains before every ethics course to talk through these very important issues so there is little or no room for misunderstanding. Generally, the up-front effort has paid off in minimizing the potential for difficulty.

That is why I feel a thorough discussion of chapters 3 and 4 is key to the ethics education experience being a valid and rewarding one. As a result, you will probably avoid some of the pitfalls I have encountered along the way.

Reflections

Prior to reading chapters 3 and 4, what do you think the role of the teacher/instructor should be?

Would the role of an ethics instructor at a police or corrections academy be different from the role of an ethics professor at a college or university?

Are you surprised that a teacher like me would feel the need to wrestle with what is okay in an ethics class and what is not?

Is teaching ethics something like teaching mathematics? Or, sociology? History? What about political science?

What would be the primary responsibilities of a teacher in an ethics course? What would be the primary responsibilities of the student?

Now you have had this discussion and thought deeply about the questions, read the next two chapters and see if your opinions change at all.

THE DEBATE OVER ETHICS EDUCATION

It may be clear to you from the first chapter that I believe a good case can be made for the wisdom of including ethics courses in justice training programs. This view itself, however, bears some examination, for there are some strong and articulate voices that express reservations about the usefulness of including ethics education in any kind of professional training program or academic course. In the interests of honest enquiry, where all views, values, ideas, and opinions are subject to scrutiny, and in the interests of modelling what is being asked of you, the reader, it makes sense to include you in a debate about the usefulness of ethics education in preparation for work in the justice field. Keep in mind that what follows is a cursory glance at what various people have to say about this topic. The intent is to offer something that will make you begin thinking about the whole area of ethics education, with the hope that you will follow up with more reading and research of your own as you seek to make your opinions and views informed ones.

THE "TEACHING" OF ETHICS

Perhaps an appropriate starting point to the discussion would be to consider what we mean by the term "**teaching ethics**"? Can ethics be taught?

"Teaching ethics" is an unfortunate term, and one that is likely to lead to considerable misunderstanding. Any teacher who teaches ethics, in the strictest sense, would presumably have to be an ethics expert; she would have to know the right answer to all kinds of moral dilemmas, including dilemmas faced by you and me as individuals. Clearly, it would be a foolish person who would claim to possess this kind of all-encompassing wisdom. Such an expert would also have to know, presumably, which of the tools discussed in Chapter 8 would invariably be successful in enabling us to make "correct" moral decisions. Few people would be likely to lay claim to this kind of expertise. Indeed, it is quite probable that those coming closest to legitimately being able to make such a claim would be the most humble and self-effacing about their expertise. Have you ever noticed how those who are most wise are often the first to question their wisdom, and those who are not wise are often the first to lay claim to wisdom? So, if there is one thing we can probably agree upon in this text, it is that such absolute wisdom is most unlikely to exist. It may be more appropriate to think in terms of a teacher or trainer *facilitating* the ethics educational experience, rather than *teaching* ethics.

Reflections
What do you think about the idea of someone "teaching" you ethics?

John Chandler (1987), professor of ethics, in Martinson's *Discourse on Ethics,* talks about reflecting on his role as an ethics professor and his pondering Socrates's question

"Can virtue be taught?" He makes a clear distinction between teaching and learning, and makes the point that "surely, morality and ethics can be learned....Whether morality can be taught discursively arouses some skepticism in all of us, but certainly there are institutional arrangements by which values can be learned" (p. 9). These "institutional arrangements" would presumably include classes facilitated, if not taught in the traditional sense, by professors. Chandler does not fully explain what he means. In any event, I think the distinction between teaching and learning is a useful one. So, rather than talking in terms of teachers and trainers "teaching ethics," it may be better to talk in terms of teachers providing opportunities for you and other students to think and learn for *yourselves* about ethical decision-making.

One of my students, at the end of an ethics course, wrote in a journal, "What you are doing is helping us understand the practical dimensions of ethics in general and how they relate to specific correctional issues. You are leading the horses (us) to ethical waters. Whether we drink or not is up to us individually."

Another wrote, "I can honestly say that I was quite surprised that you had not overtly attempted (as far as I could tell) to influence our thinking. I thought you would attempt to try this in a way that was discreet and casual."

These comments reflect the students' experience of an ethics course. In other words, it is important that there always be scope and safety in the classroom for personal choice and personal decision-making.

THE PURPOSE OF TEACHING ETHICS

What is the purpose of activities such as discussing and debating ethical issues? Would it be reasonable to assume that if one works in an environment characterized by a "subculture" (discussed briefly in the introductory chapter and discussed in greater depth in chapters 5 and 6), there may be considerable pressure upon an individual to follow the norms of the group in which she works, and to do the accepted thing? If individuals choose not to yield to the subcultural pressures, the penalties can be severe. That being the case, would it again be reasonable to assume (and this is borne out by my research, discussed in the chapters mentioned above) that individuals from time to time may choose to act less ethically than they might do otherwise, simply because they consider the potential costs of doing what they think is the right thing to be too great?

What, then, are justice educators to do, particularly those who are concerned with the ethical aspects of working in corrections or policing? Should they simply inform you, the student, that this is the way things are, and that the subcultural constraints are a reality that has to be accommodated if you are to survive? Should they ignore this facet of work in this field and allow you to learn about the constraints the hard way? It is my experience that the more common approach is the latter one. Indeed, a quick, informal survey of police education and correctional worker students, and of police officers and correctional officers, indicated that the issue of subcultural constraints generally is not addressed in training programs. Some of these respondents, after I addressed the issue in ethics seminars they were attending, expressed concern in some cases, and anger in others, that this was the case.

Often I have asked my own students what we should do. Would they prefer to be told about the justice field "warts and all," or would they prefer to receive a kind of "laundered" version—one that is glamorous and gussied up? Almost without exception the response is that instructors should tell it like it is. Not only should we do so, say the students, but to fail to do so would be improper and even unethical.

The risk is that you, the student or trainee, may develop a distorted view of the field. You may put your own spin on the discussion, lose some of your early enthusiasm for the officer's job, or approach the job more cautiously and defensively than is warranted.

The clear message I receive, however, is that students want those in the know about police and correctional officer roles to be up front about the good and bad sides of the job, the advantages and the disadvantages; and that the risks of not being absolutely honest about all aspects of the job outweigh those of being honest.

Reflections

What is your take on this issue? Do you feel that your teachers should tell you about the potentially unpleasant and troubling aspects of working in the justice field?

Should they attempt to heighten your awareness about group pressure and how it can impact on your sense of individual moral responsibility?

If you feel they should, is there any risk that this information may unfairly colour your perception of the job, or lead you toward cynicism even before you get started in your career?

If you feel they should not, and that it is better for you to find out about the troubling parts of the job on your own, might you in some way blame them later on for not fully informing you?

IS THERE A PLACE FOR ETHICS EDUCATION?

Let's return to the original question: Is there a place for ethics education in justice training? As was stated earlier, there is quite a debate surrounding this issue and there are some equally convincing voices on both sides.

A report issued by the University of Alberta Task Force on Ethics (1985) and a subsequent Progress Review (1988) provide a useful framework for this discussion. They serve as catalysts for examining some of the various perspectives on ethics education. The original report was based on the results of an extensive survey of opinions of faculty, students, and professional associations. The survey results led the university to commit itself to the idea that it should play an active role in ethics education (p. 67). Indeed, the first recommendation called for a required exposure to ethics education for all first-year undergraduate students:

> That the University require all first-year undergraduate students to take a course with an ethical component. This need not mean that every student must take one particular course, but may involve a choice from a variety of courses which have been identified as having a significant ethical dimension (p. 73).

Most important, the report makes it clear that its purpose "is not to promote the instilling of specific ethics, but to raise the level of awareness about ethical issues" (p. 24). This is a critical distinction to make, and it helps us develop a better understanding of the difference between teaching ethics *per se*, and providing an opportunity for learning about ethics and moral decision-making. If I regard my role as one of "teaching ethics," then, as we said earlier, this implies I know best, in which case I would probably wish you to do what I suggest, or tell you to do. It also raises the ugly possibility of what I call the lurking dangers of indoctrination. This topic is discussed in greater detail in Chapter 4 (see p. 47).

The university survey results are interesting. Eighty-two percent of 1 025 students surveyed agreed that a "university education should help a student develop ethically" (p. 58). The professors' responses were recorded by department, the highest rate of agreement to this concept being found in the faculties of law (91 percent) and health sciences (89 percent). This is encouraging for those of us who feel that one day we may need the services of either a lawyer or a doctor! Or, might it indicate that these are the two professions most in need of ethical attention?

The Opinions Are Split

As indicated earlier, and not surprisingly perhaps, there was quite a diversity of opinion expressed by the university professors in their written responses to the survey questions. More often than not, their comments were either completely *for* the teaching of ethics, or they were completely *against* it. This is a topic that seems to attract strong opinions in one direction or the other.

The comments of those opposed to the notion of ethics education are particularly interesting. One respondent noted, "We must be careful about a university trying to shape ethics and morality for the community as a whole, because different people have different conceptions of values and ethics" (pp. 37–38). Another responded, "I don't believe it [the teaching of ethics] should be a goal in itself. The ethics systems of our university-age students are already in place and are basically unalterable at that age" (p. 38). One response was clearly conditioned by the fear of indoctrination: "University professors are, other than in intelligence, like the general population, and include some people who lack moral judgment. In God's name, no potted required courses in ethics and values! This results in brainwashing, not critical thinking, in promoting 'acceptable attitudes,' not hard thinking" (p. 38). And finally, "No one becomes ethical merely by teaching or taking a course in ethics. Every course ought to be an ethical exercise" (p. 38).

The caution expressed in the first quote is one worth heeding. As we have indicated, we should take great care not to attempt to legislate a particular concept of morality, or to claim any monopoly on wisdom.

The second sentiment, adopting the view that students' values are pretty well fixed by university age, is a surprising one. Many people who are involved in any kind of teaching activity with adults speak about the way values can change as a result of the educational process. This would be particularly true of students who may have simply "adopted" their value system from others, without thinking about those values in any real way. One of my students wrote in a journal, "Before coming into this [Ethics] class I felt that once you reached a certain stage your ethics were set, but because this course has made me look at myself, certain things have changed." And another wrote, "I have learned that my opinions and beliefs are not carved in stone, and may be changed if presented with a good argument."

The third sentiment again is one worth heeding. The legislating of ethics and values by an individual or organization necessarily forecloses individual critical thinking. The danger here would be that an individual might be required to subjugate his own sense of morality in the interests of fitting in.

The fourth and final sentiment also requires serious consideration. To think of a course in ethics as some kind of panacea for all the ills of a particular justice system would be foolhardy. The view that every facet of a justice program should include consideration of ethical issues seems a more appropriate approach.

Nonetheless, most of the opinions in the report were in support of the idea of ethics education in principle, although there was some disagreement as to what form this should take.

Reflections

Are you surprised that 82 percent of the 1 025 students surveyed for the University of Alberta report agreed that the university should help them develop ethically?

What is your view on the opinion expressed by one professor that by the time students are of post-secondary age (generally 18 and up) their values are pretty well set?

One professor is clearly of the view that ethics courses are designed to promote "'acceptable attitudes,' not hard thinking." What do you think of this point of view? (Note: This is a question that you could and should be asking yourself at this point in the course and at its conclusion).

Two writers in the ethics area, Daniel Callaghan and Sissela Bok (1980), in their discussion of the doubtful goals of teaching ethics, affirm,

> It is often said that one test of the success of teaching ethics would be change in the moral *behaviour* of students, and that a central goal should thus be an attempt to change behaviour.... this is an exceedingly dubious goal. First, even if a course could change behaviour, it is hard to see how, short of constant reinforcement of the new behaviour, its effects could be long or surely sustained once out of the classroom; other influences would play their role, and no course could be a permanent antidote against them (p. 69).

In the justice context, of course, one significant influence would be that of subcultural constraints. And these writers are in all likelihood correct in affirming that it would be highly improbable that an ethics course, in and of itself, could be a permanent antidote for those influences. My own experience of teaching ethics courses and conducting ethics seminars is that they do, indeed, have a "shelf life"; that is, that any effects begin to wear off after a while. Several students have talked to me about the importance of providing what they call "booster shots."

The discussion also raises the issue of the importance of thinking about supplementing ethics courses by encouraging teachers and students to consider the ethical implications of all elements of any educational program in justice departments. Consideration of the ethical dimensions of courses on, for example, Evidence, Arrest Techniques, Court Procedures, Security Techniques, or Use of Force Legislation, to name just a few, could build upon learning and skills developed in a foundation ethics course.

GIVING ADVICE

Is there a place for **advice-giving** by your teacher in ethics courses? There are those who feel there may be a place for advice-giving as long as it is done carefully and sensitively by a credible and knowledgeable person. One of my teachers at the University of Toronto, in fact, advocated such a viewpoint. Clive Beck (1991b), in "How Adults Learn Values," wrote,

> People have a strong interest in the outcomes of their behaviour and will adjust their values and actions in the light of information about consequences of behaviour. Much of this information can be made available through instruction (p. 14).

Michael Davis (1990), in an article entitled "Who Can Teach Workplace Ethics?," takes a similar view. One way of teaching ethics, he claims, is what he calls the "prudence" approach:

> You explain right and wrong in terms of what the employer, the law, or the profession wants and what will happen if one disobeys. You might, for example, explain why an employee should be prompt in this way: "If you don't want to get fired, arrive on time" (p. 30).

There may indeed be a place for this kind of prescriptive approach in justice training, because someone who knows the expectations and fully articulates them can save a new officer from considerable grief later. It is particularly important to pass this information on to recruits, because only then can they reasonably be held accountable for subsequent unacceptable behaviour. It would be important here, of course, to inform students about policies and procedures, the mission statement if there is one, and the "lore" of the institution or the force—the unwritten rules of conduct governing correctional and police officers.

This kind of advice-giving about the rules and expectations of a particular workplace by teachers, however, as important as it is, is not the same as teachers telling you how you should respond to ethical dilemmas you encounter in the course of your job.

THE COLLABORATIVE PROCESS

Earlier we talked about the role of the teacher in facilitating the ethics educational experience. While there may be a place for teachers prescribing employer expectations in terms of behaviour, it would be more consistent with a facilitative approach to learning for the teacher and student to engage in a collaborative process. This approach would be characterized by the teacher being both a teacher and a learner, and the student also being both a teacher and a learner. Clive Beck (1991a), in "Approaches to Values Education," writes,

> a large proportion of the teaching is done by the learners, who instruct each other and their "teacher." The main point of values instruction, therefore, is not that one person tells other people what to value, but rather that there is explicit, systematic, joint enquiry into value matters (p. 8).

In the business context, Dan Rice and Craig Dreilinger (1990) also favour the collaborative approach. The ethics experience, they affirm, should accomplish three goals: 1) heighten employees' ethical awareness; 2) provide them with the means to identify ethical issues; and 3) help them develop tools to clarify and resolve ethical issues (p. 106). There is no hint of "telling" or prescriptiveness in this view; the approach is one characterized by cooperation and helping on the part of the "teacher."

One of the reasons we should be careful about an overly prescriptive approach to ethics education is that as human beings we are each unique. Therefore, each of us necessarily brings to any moral dilemma subjective perspectives. Nel Noddings (1984), in her book *Caring: A Feminine Approach to Ethics and Moral Education,* talks about the "uniqueness of human encounters including those in the moral domain" (p. 5). She adds,

> Since so much depends on the subjective experience of those involved in ethical encounters, conditions are rarely "sufficiently similar for me to declare that you must do what I must do" (p. 5).

And Daniel Callaghan and Sissela Bok (1980), whom we mentioned earlier, in a book they edited called *Ethics Teaching in Higher Education*, also caution against a "telling" approach:

> No teacher of ethics can assume that he or she has such a solid grasp on the nature of morality as to pretend to know what finally counts as good moral conduct....It is the time and place to teach them [students] intellectual independence, and to instill in them a spirit of critical enquiry (p. 71).

Reflections

What are your views about having a separate "foundation" course in ethics, supported by you and your teachers, considering the ethical dimensions and implications of all your program courses?

Is there a place for advice-giving by your teacher in the context of an ethics course? If so, what kind of advice would you a) expect to be given, and b) feel okay about being given? If you feel there is no place for advice-giving by the teacher, why not?

In the context of a justice program, do you think there is a place for a collaborative approach to learning between you and your teacher?

SUMMARY

We have discussed whether there is a place for ethics education in justice training. We have suggested that because of police and corrections subcultural constraints on the autonomy of the individual, for these people, taking time out for thought and reflection about ethical issues and moral dilemmas might be an important and wise use of time. We have suggested that stand-alone ethics courses lose much of their potential effectiveness if attention is not also paid to the ethical implications of all aspects of the justice training curriculum. We have suggested that the role of the teacher is an important one, but that how the role is played deserves some attention. We have suggested that there may be a place for advice-giving on the part of the knowledgeable and experienced teacher, but that this advice-giving should be engaged in cautiously. And finally, we have suggested that a collaborative approach to learning in an ethics course might be the least problematic approach in terms of negating the potential effects of teacher influence.

In the chapter following, we will consider this latter point in greater detail.

chapter four

The Role of the Ethics Educator: The Lurking Dangers of Indoctrination

I sometimes found your mannerism was downright frustrating, because you did not help us make right choices.... You helped prepare us for possible situations we might encounter, and the dilemmas we might face in those encounters.
                    —Journal entry. Justice student in ethics course.

During the course a lot was said about teaching ethics and being influential. I don't think a person can teach ethics and not be influential.
                    —Journal entry. Justice student in ethics course.

I believe that a teacher has just as much of a right in stating his or her opinion and thoughts on any issue, as long as the students are aware that all people have this right.
                    —Journal entry. Justice student in ethics course.

I was sure that this course was meant to change my opinions and my beliefs. I also thought that you, the professor, thought your ethics were the correct ones, and planned to force them on us. However, something changed along the way and I began to realize there are two sides to every issue. Maybe my opinion wasn't always the right one.
                    —Journal entry. Justice student in ethics course.

I don't think that the teacher influences *what* I think, but I do think he influences *how* I think.

—Journal entry. Justice student in ethics course.

CHAPTER OBJECTIVES

After reading this chapter, you should be able to

- Identify and explain some dangers with respect to the role of the ethics educator
- Describe the notion of **teacher influence**
- Explain why the teacher-student dynamic is a particularly awkward one in the context of ethics education
- Identify at least one reason why students sometimes do not trust a teacher's invitation for them to think critically and independently
- Identify four characteristics of critical thinkers

THE WORRY ABOUT INDOCTRINATION

You may wonder why, in a book that has been written for students like you, I have included a discussion on the role of the ethics educator. Let me explain.

Many ethics educators worry about their potential for becoming doctrinaire in their approach to ethics education or for being perceived by their students to be so. This is a legitimate concern. That what I do in the ethics classroom may be perceived by my students as indoctrinatory has caused me some angst over the last 20 years since I started teaching ethics courses. The word *angst* is used advisedly. Much of the discussion that follows is based on my own personal experience and struggle. My hope is that this discussion and debate will provide interesting and provocative food for thought in your classroom. I would suggest that you think deeply about this important topic, and I would also encourage you to be as honest about it as you can be in any classroom discussion.

Clearly, my discomfort would seem to indicate that I see the words *indoctrinate* or *doctrinaire* as necessarily negative ones. I consider a doctrinaire approach one to be avoided studiously. Perhaps I should explain here, without getting too deeply into philosophical discussion, what I mean by "indoctrinate." What I mean by the notion, first, is the attempt by one person (teacher in this case) to get another person or persons (students) to accept a particular position. Included in the notion, second, would be the intention by the teacher to withhold from a student the opportunity and right to consider that position for herself, and to develop her own perspective on it.

Instinctively I feel that there is something improper or inappropriate about one adult, especially one in a position of power as would be the case with a teacher, imposing on adults his views on various moral issues.

However, I have to confess that a significant problem for me is that I tend to hold strong personal views on some of the moral issues I discuss with my students. That is not to say I am inflexible; indeed, over the years I have come to accept different perspectives on many of the issues discussed, because of the contribution of my students. On other issues, however, nothing that has been said by others has influenced my thinking. That being the case, it is a source of concern for me that some of my stronger views may in some way "leak out" during discussions with students.

SHOULD AN ETHICS EDUCATOR COME CLEAN ABOUT HER VALUES?

It should be noted, of course, that some ethicists and educators feel that laying one's own views on the table is an appropriate and necessary thing to do. The knack, they claim, is to ensure that one's students are made to feel no pressure to accept those views and are encouraged to express alternative ones. Indeed, some are of the view that to refrain from doing so may be dishonest or manipulative.

A philosophy professor, during one of my postgraduate courses at university, used this technique of stating his views on each topic prior to class discussion. As a student, I was taken aback when he led off each discussion by arguing forcibly for his own view. I asked him whether he felt there was any danger of his foreclosing dissenting student views because, as the professor and the awarder of grades, he might intimidate students. Did he feel students might feel obliged to acquiesce to his point of view? His response was that it was his responsibility to "come clean" about his own views, put them on the table, get them out in the open—but then work to create an atmosphere where students would feel comfortable in attacking those views if they could.

I am not entirely convinced. There! I have expressed a personal opinion on this issue. There is a distinct danger, in my view, whether one is a professor or a staff trainer at a police or corrections training academy, of being perceived to be overly influential on the group. The teacher is invariably going to be seen by the students, and with some justification, to be in a position of power, and this is going to be an ever-present dynamic in the classroom. This potential for influence may be less of a problem at the postgraduate level, but it is still there; and at lower levels of education it is very much there.

During one class with this particular professor, I gathered up my courage and expressed surprise at his approach. Surely, it is a risky activity to argue his own views? Surely, it would be more appropriate to attempt to minimize, in whatever way he can, his potential for influencing his students? During the discussion, one classmate gently reminded me that, as laudable as it was that I was concerned about the dangers of too much influence over the thinking of others, it was inevitable. At another time, it was pointed out to me that the fact that I am male, White, speak with a British accent, and may be perceived by students as something of an expert, means I can operate from no position other than one of power. Try as a teacher might, therefore, it can be extremely difficult to create a climate of open, honest debate with his students, whether he comes clean about his views or not.

Sick Days

Perhaps at this point it would help the dialogue if we were to examine one of the ethical issues I discuss with my students. The issue is whether it is ethical for an employee to take a "sick day" when he is not sick. This is a particularly contentious issue that inspires some lively discussion in my classes. It is also an issue about which I hold some strong views.

Simply put, I feel it is unethical for employees to take a sick day for any reason other than sickness. This is *my* value and it governs my behaviour with respect to this issue. This is not to say that I am correct or that I have not and do not consider other perspectives. Frankly, however, my belief is that for me to take a sick day when I am not sick is tantamount to theft. And, other than the obvious legal difference, I see little or no moral difference between taking a sick day when I am well and helping myself out of petty cash to whatever I am paid for a day's work.

Students tend to become defensive when we discuss this issue. They often engage in what I consider to be rationalizing. Sick days, they argue, are provided for in the employment contract and they are an employee's right. Who is to say what sickness is or what it is not? What if, despite my not being sick, I come back to work the following day and perform better as a result? All kinds of what I consider to be red herrings are drawn across the trail. Participants tend to struggle for a reasonable and defensible response when it is suggested to them that such provisions in the contract are there in case there is a genuine need to use them; that the sick day provision is made for *bona fide* cases of sickness.

Influence by Accident?

The challenge for me—and I constantly question this—is whether, given my strong feelings on this issue, and even though I do not express an opinion, it is possible for me to intellectually challenge others' ideas and perspectives in a dispassionate way. *It is this sort of question that is key to worries about indoctrination.* I am not only a person holding power, but a person holding strong views on certain important issues. We need to ask ourselves whether these are, potentially if not actually, two active ingredients in the indoctrination process.

What if, hypothetically, my views can be demonstrated to be morally sound? Does this give me the right to influence or attempt to influence the views of others, in other words, to have their thinking become more akin to mine? Is the role of the ethics educator one of deliberately setting out to make individuals more ethical, to have them think and act in a way prescribed by the educator? Or is that role one of mutual engagement in a process of critical thinking about, and examination of, issues?

Fertile Ground for Influence

Let's say for the moment that the latter question reflects more accurately what should be going on in an ethics classroom; that you, the students, are encouraged to think critically and to think for yourselves. Might there, however, still be a possibility of the ethics educator, especially one who is well-liked by you, seducing you into thinking in a given way? If the learning climate is a comfortable one, where the teacher is respected, trusted and well-liked, might this be fertile ground for the overt or covert manipulation of your group to that teacher's own views? Let's consider another example.

Later in the book, reference is made to a conversation I had with a student when I was administrator of the justice department at the college where I work. The student was in the police education program and he had been caught cheating on an examination. When I asked him how he would square this activity of cheating with the trust and responsibility society would invest in him as a police officer, his reply was "It would be different if I were in uniform, I would be honest on the job." Several years later I am still trying to come to terms with this response and the kind of thinking and morality that would inspire it. The student clearly felt that it was okay to be dishonest in one's personal life, but not in one's professional life. There appears to be little or no awareness of the possibility that the two may be connected.

This incident is often a topic of discussion in my classes. It is common for participants to defend this student's response. What we have here, then, is a situation where, in a real sense, I would like the participants to see that there may be something worrying about the student's response—in other words, to come around to my way of thinking. As a result of the debate and the concerns I raise, some may change their thinking and come to see that perhaps there *is* something worrying about the student's response.

What has happened here? Have I been guilty of indoctrination? Or have I simply been an agent of change for those who have decided for themselves to embrace the change? Certainly, no one is prohibited from, or judged for, continuing to think in any way she chooses on the issue. The students are challenged intellectually, and are asked to offer reasons for their views. And even if they can't defend their views, they are still, of course, free to continue holding them, unreasoned and contradictory as they may appear to be.

Comfort can be taken from the fact that wishing students would embrace a view I hold strongly may not be the same as inducing that embrace by way of psychological coercion or by overt or covert suppression of different ways of thinking. But the issues of power and control on the part of the teacher remain, and in my view need to be constantly guarded against. Sociologist Kimberley Folse (1991), in a journal article entitled "Ethics and the Profession: Graduate Student Training," refers to Churchill (1982), who makes the point that "moral values cannot fail to be taught" by teachers (p. 306). If this view has any merit, teachers find themselves in an unenviable position, through no fault of their own. It is as though they have had this (unwanted) mantle of responsibility placed upon them.

It seems to me, from the many conversations I have had with my students on this topic, that you are going to place this mantle of responsibility on us as teachers, no matter how much we protest. My students tell me that despite my pleading that the last thing I wish to do is influence them, this is exactly what is going to happen. First, they tell me that I am older, I am better educated—how could I not influence them? Second, they say, if you throw respect-for-the-teacher into the mix, the potential for influence is magnified several times. Perhaps the ethics teacher who would be safest in this regard is the one who is not much older than the students, who is not well-educated, and who is not well-liked! Of course, I am being facetious here, but you can see what I am getting at.

As much as I am discomfited by the thought that I am going to be influential, whether I like it or not, my students make a convincing case. It is important that you discuss this important issue with your teacher or instructor. The following Reflections may help.

Reflections

Should a teacher, particularly in an ethics course, be worried about being—or being perceived by his students to be—doctrinaire?

Would you prefer your ethics teacher to "come clean" about her personal views on a topic under discussion, or attempt to conceal them?

With respect to guarding against teacher influence, what do you think is the responsibility of the ethics teacher when one student says in class, "I think the trouble with this country is Chinamen and Pakis [sic]," or another says, "Aboriginal people are given too much money by this government, and they are all a bunch of drunken bums?"

Is there a place in the classroom for teacher influence if it is exercised in the challenging of racist or sexist remarks?

To what extent, if any, should teachers attempt to influence students who exhibit socially unacceptable attitudes?

If you think there is a place for a teacher to challenge socially unacceptable opinions, might this get in the way of the teacher being able to create an open climate where students feel they can state what they truly feel on a given topic?

What is your experience with respect to teachers and how they can influence you?

THE TEACHER-STUDENT DYNAMIC IN ETHICS EDUCATION

As we have suggested, because of their role, teachers necessarily find themselves in positions of power. They are in a unique position to influence. The potential for influence generally has its roots in either the student's fear of, or respect for, the teacher, or simply in the fact that how well the student does in the course is perceived by the student to be in the teacher's hands.

The teacher–student dynamic is a particularly awkward one in the context of ethics education because of the natural propensity of students to see teachers of any subject as experts. Often the relationship, somewhat sadly, is characterized by the student seeking to "give back to the teacher what the teacher wants." Sometimes this may stem from the student's respect for the teacher's expertise, but more often it stems from the view that feeding back to the teacher what the teacher has given out is the best way to ensure a good grade. One of the more disappointing moments for me in working with students occurs when someone approaches me to find out what I "want" on an examination or in an essay. Such enquiries indicate that I have been unable to convey to the student that what I am really interested in is what *they* are thinking about a given topic.

It is a significant challenge, in my experience, to create a safe climate in a classroom where students feel they do not have to worry about agendas—including hidden ones—that the teacher may or may not have. In an effort to combat this fear, I commence any ethics seminar by informing the students that I make an honest attempt to harbour no agenda, hidden or otherwise; I attempt to convey to them that the classroom is a safe forum for the sharing of ideas, and that there are probably going to be no right or wrong answers regarding the issues raised. In other words, the process of airing and exploring ideas is the important thing.

TRADITIONAL EDUCATION AND STUDENT CONDITIONING

The problem is that this kind of educational approach in which students are encouraged to think for themselves generally flies in the face of virtually all their previous educational experience. It is such an atypical educational approach that some become suspicious of it, and others even intimidated by it. They have become so used to seeking and finding answers in the classroom that even if one asserts this is not part of the process here, they feel that this must be what the teacher is about. Often they feel, probably because of the conditioning that has gone on in many classrooms during the previous 15 years or so, that in a subterraneous or devious manner, the intention is to attempt to lead them to think in a particular way. Often, attempts to disabuse them of this notion fall on deaf ears or are met with a reluctance to accept what the teacher is saying.

Leo Buscaglia, the American university teacher who has made a name for himself talking about "love," graphically depicts the problem with teacher- centered education in the foreword to his book *Love*. "Education," he writes, "should be the process of helping everyone discover his uniqueness.... But we still see the processes again and again of trying to make everyone like everybody else." He continues,

> A few years ago with some of my student teachers at the University, I went back into classrooms and was astounded to find the same things going on that had been going on when I was in school—a million years ago. For example, the art teacher would come in. Remember how we always anticipated and got ready for the art teacher? You put your papers down and you got your Crayolas out and you waited and finally in would walk this harried person. I really feel sorry for an itinerant art teacher. She comes racing in from another class and has time only to nod to the teacher, turn around, and say, "Boys and girls, today we are going to draw a tree." She goes to the blackboard, and she draws *her* tree which is a big green ball with a little brown base. Remember those lollipop trees? I never saw a tree that looked like that in my life, but she put it up there, and she says, "All right, boys and girls, draw." Everybody gets busy and draws.
>
> If you have any sense, even at that early age, you realize that what she really wanted was for you to draw *her* tree, because the closer you got to her tree, the better your grade (pp. 20–21).

Buscaglia goes on to express concern about this tendency to demand conformity in the classroom, which, he says, continues "right on into the university." "We don't say to people, 'Fly, think for yourselves.' We give them our old knowledge, and we say to them, 'Now this is what is essential. This is what is important.'"

If Buscaglia is correct, educators should not be surprised if students come into the classroom thinking that academic success depends on the degree to which they conform to the teacher's expectations. Presumably, this requirement to conform would apply to an even greater extent in police and corrections training academies because of their paramilitary nature.

IS IT POSSIBLE TO OVERCOME STUDENT CONDITIONING?

My experience is that there are ways to minimize these suspicions, but I have found that it is not possible to eradicate them. Of course, there are always students who, for whatever reason, seem to be more or less amenable to the new approach. Then there are those who, despite one's protests, continue to feel that the purpose of the exercise is to get them to think in prescribed ways; in other words, that the real purpose is to indoctrinate them.

For example, one of my students wrote,

> During the course a lot was said about teaching ethics and being influential. I don't think a person can teach ethics and not be influential.

And another student, in response to the different, more open, approach to learning, said,

> There is no way that a lifetime of conditioning [concerning previous classroom experiences] can be changed within a semester....

Perhaps a more sobering comment is this one:

> You are well-liked and held with great respect. I feel that people were afraid to say what they felt because they didn't want to lose your respect. In the back of our minds we knew this was wrong and you would never do it, but because of our conditioning through school, that was always in the back of our minds.

The latter comment is one that gives the educator cause for concern, but it also provides, in a reverse way, a glimmer of optimism. The student who wrote the comment a) was able to express the concern, and b) felt free enough to sign her name.

Here are some other student journal comments for you to think about:

> If the teacher did put his opinion into the conversation, I feel it would be a bad thing because some students would conform to his values and beliefs.

> There was never a time when the teacher was overly influential. At times he seemed to be concerned by the way certain individuals were thinking, but he never really judged us as unethical. Overall, I think the teacher was influential to the extent that that's what teachers are for—to be of some influence, to help us learn.

> If there was any time that the teacher was influential upon students, it was in their best interests. I think this is all right because he knows what we are going to be getting into and has already experienced it firsthand. He is only trying to prepare us to be successful.

> I was sure that this course was meant to change my opinions and my beliefs. I also thought that you, the professor, thought that your ethics were the correct ones, and that you planned to force your ethics on us. However, something changed along the way and I began to realize there are two sides to every issue.

Reflections

It would seem from these comments that despite my protests, I was only half successful in attempting to neutralize my influence over the students. What do you now think about this awkward and somewhat complex issue of teacher influence?

THE IMPORTANCE OF LOOKING AT ALL SIDES

On the surface, what I seek to achieve in my interaction with students is to open their minds to alternative points of view on a given topic, such that they are required to consider many sides of the issue before coming to a conclusion. Richard Paul (1992), who is a North American expert on critical thinking, in a chapter entitled "Ethics Without Indoctrination," argues that ethics can and ought to be taught, "but only in conjunction with critical thinking" (p. 240). He goes on to assert that in order to teach ethics morally, the course should be designed in such a way that students become "as adept in using critical thinking principles in the moral domain as we expect them to be in the scientific and social domains of learning" (p. 243). I agree. In my courses I spend time covering some of the principles and skills of critical thinking. I do so for two reasons: 1) I agree with Paul that it would be unethical to attempt to teach ethics to individuals when they are unlikely to have the capacity or permission to sort through the arguments themselves, and 2) it is a small but significant step toward guarding against the lurking dangers of indoctrination or the perception of indoctrination. The teaching of critical thinking, however, is more successful in protecting against actual indoctrination by the teacher than it is in protecting against the *perception* of indoctrination. Almost invariably, in my view, there will be students who, despite the teacher's best efforts, will become or will continue to be suspicious of the ethics educator's motives.

Here are a few brief comments made by three of my students about the importance of critical thinking. They are included for discussion purposes. The first one calls into question the idea that the educational process should be one that requires the teacher, as expert, to provide facts and information, and the student, in order to be successful, to simply regurgitate this material in examinations or papers:

> I have always had the notion that we never truly think while we are in college. All we do is memorize.

The second comment, even though it is expressed in the way that baseball manager Yogi Berra might have expressed it, contains an idea worth heeding:

> I did find the critical thinking part quite interesting, because it goes to show how much thought we don't put into our thinking.

And last,

> It is easy to accept the *status quo* so as to not rock the boat. However, if throughout history no one thought critically about various policies and issues, and had not the courage to ask "why?", we might still be in the Dark Ages.

Indeed, had no one had the courage to think critically, perhaps we would still be convinced that the earth was flat, or that it was impossible for humans to fly.

DIALOGUE

Let's return to the indoctrination worry. Clive Beck (1991c), one of my teachers referred to earlier, stresses the importance of dialogue in ethics education, and he sees this approach as one way of avoiding the possibility of the teacher being accused of being doctrinaire.

He does caution, however, that "Even with the dialogue format, authoritarianism may continue" (p. 3). He continues, "Careful planning and constant vigilance are needed to ensure that learners have freedom of speech, belief, and action and that the influence is reciprocal" (p. 3). Borrowing from Freire (1988) and Habermas (quoted in White, 1988), he lists the key elements of dialogue:

- Respect for each other's insights
- Respect for each other's "tradition" or "story"
- Shared control of the form and content of dialogue
- Focus on concrete, lived experience
- Testing through action (p. 2)

Noted sociologist Max Weber (1967), however, would probably argue that this vigilance against the dangers of indoctrination would not be a sufficient safeguard. He was writing in the context of science education, but one can only assume that his forebodings would be even greater in the ethics education context. He warns of the perils of teachers taking a stand on issues. For example, of political debate, he says,

> They [words] are not ploughshares to loosen the soil of contemplative thought; they are swords against the enemies: such words are weapons. It would be an outrage, however, to use words in this fashion in a lecture or the lecture room....But the true teacher will beware of imposing from the platform any political position upon the student (pp. 145–146).

He becomes almost strident in warning teachers, in their professional role, against attempting to answer, or being drawn into, what he calls "questions of the value of culture" (p. 146). "The task of the teacher," he affirms, "is to serve the students with his knowledge and scientific experience, and not to imprint upon them his personal views" (p. 146).

OTHER VIEWS ON THE EDUCATOR'S ROLE

Two writers and researchers who have written extensively on ethics, Callaghan and Bok (1980), have some interesting things to say about this subject. Ethics courses, and by implication ethics teachers, they claim, "should make a change in the way students *think* about ethical issues...by providing them with the tools for a more articulate and consistent means of justifying their moral judgements" (p. 70). One of these tools would be the ability to think clearly, constructively and well. (A discussion of this particular tool and others follows in Chapter 8).

And Jonathon Baron (1990), in a journal article called "Thinking About Consequences," makes an interesting point about the importance of moral thinking:

> Much of what we call immoral behaviour, such as the behaviour of Hitler's followers, or less insidiously, the followers of Colonel Oliver North, arose not so much out of malice but rather out of *thoughtlessness*, that is, *failure to think of consequences and relevant moral principles, or out of unreflective commitment to questionable ideologies* (p. 77). [emphasis is mine].

Baron believes not only that there is a place for moral education, but that it is absolutely vital if we are going to develop students with the capacity to think for themselves about ethical issues and moral dilemmas encountered in their personal and professional lives.

One of the ethics teacher's responsibilities would therefore be to ensure that her students are encouraged to do their own critical thinking. It is important that students understand clearly the differences between critical and uncritical thinking. For example, I advise my students that critical thinkers

- tend not to jump to conclusions;
- examine the credibility of the evidence presented;
- are humble and fair-minded; and
- courageously face up to, and incorporate into their thinking, alternative points of view, if, after examination, these appear to be credible.

Uncritical thinkers, I tell them, tend to do the opposite.

How would you describe your educational experience up until now? Have you, generally, been encouraged by your teachers to think for yourself?

Reflections

What do you think about the differences between critical and uncritical thinkers? Where do you see yourself in this respect?

How do you respond to the views quoted from the journals of my students?

Would you agree, based on the student comments, that I was only half successful in my attempt to neutralize the potential for influencing my students?

What might be the implications of this partial failure on my part for your experience in an ethics course?

Based on the brief description of critical thinking above, do you consider yourself to be a critical or uncritical thinker?

SUMMARY

How can we sum up this discussion about the role of the ethics educator?

We have discussed the role of the ethics educator and we have considered the worry about a doctrinaire approach to ethics education. We have discussed whether it is wise, or appropriate, for the ethics educator to "come clean" about his own views on a discussion topic. We have raised the question of whether it is possible for a teacher who chooses not to express her opinions to conduct a dialogue with students in a dispassionate way. Also raised was the dilemma for the teacher in establishing a safe, open climate for learning, while at the same time challenging socially unacceptable values. We also discussed the challenge that the ethics educator faces in creating a safe learning climate of independent and critical thinking, when most students' educational conditioning is characterized by the notions "teacher knows best" or "teacher as expert."

I wish you well, whether you are an instructor or student, in your ethics educational experience. I feel confident that an honest, open discussion about the foregoing points will help to head off any unpleasant misunderstandings between teacher and students, or, indeed, between student and student.

Try to be as moderate and understanding as you can be, especially at those moments when there is the potential for heated debate. Keep reminding yourself of the peculiar and inherent *awkwardness* of discussing personal values in an open forum, particularly when those dearly held values may be put to a severe intellectual, and perhaps emotional, test.

In conclusion, the *process* of an ethics educational experience is such an important one that it is worth any investment of time and discussion to avoid the pitfalls. I wish you well as you seek to clarify your thinking and feelings on very important issues. I have little doubt that you will be a more complete and prepared justice professional as a result of your effort.

chapter five

Subculture and the Individual Officer

So you either fit in, or you don't fit in. If you don't fit in you are not going to last. It's the same as the criminal subculture.

—Justice student after field practicum.

One thing that really scared me was the need for acceptance within the group, which often leads to betraying your own sense of self, and the values we hold.

—Justice student after field practicum.

The big dilemma was, do you stand alone and become a leper or do you let peer pressure and group think compromise your ethical stand?

—Justice student after field practicum.

There is a price tag attached to honesty, especially in corrections. That is a scary thought. That even though you may be right, and that the actions of a fellow employee may have been wrong, in a subculture it is better to button up than to open up and speak to someone and risk being an outcast.

—Justice student after field practicum.

I know I have spoken out in the past at work and I've suffered the consequences. Most times now I just keep quiet.

> —Journal entry. Professional person in ethics class.

I can remember a specific topic from the class: "Does silence imply consent?" Maybe there is another question that should be considered: "Does silence imply fear?"

> —Journal entry. Justice student in ethics class.

CHAPTER OBJECTIVES

After reading this chapter, you should be able to

- Explain the significance of including a discussion of subculture in a book about ethics
- Explain what we mean by the terms **subculture** and **subcultural constraints**
- List six norms that characterize police and correctional officer subculture
- Identify two of the more significant norms in police and correctional officer subculture
- Explain why police and corrections subcultural norms may be more compelling to officers than subcultural norms in most other workplaces

THE IDEA OF POLICE AND CORRECTIONAL OFFICER SUBCULTURE

As you will have noticed, the first three chapters referred to the idea of police and correctional officer subculture. I have also alluded to the idea that subcultural constraints may inhibit individual officers from making their own independent, moral decisions. It is because of this latter sentiment that I believe any text addressing ethical issues in the justice field would be incomplete without also addressing one of the more constraining forces against individual moral decision-making—namely, that of a subculture.

In this chapter we will more closely examine the idea of a subculture as it pertains to police and correctional officers. What is a *subculture*? Is there a difference between a subculture in the justice context and a subculture in other work contexts? What are its defining characteristics?

It is important to note that police and correctional officer subcultures are not entirely similar, but I would suggest that there are enough similarities, including the crucial one of colleague loyalty, to consider them together. Clearly, the differences in role between the police officer and the correctional officer would affect the individual's perception of his group, and, indeed, how his group's norms are established. For example, each group works on a different side of the fence, so to speak. Police officers apprehend offenders and correctional officers work with offenders toward their release back into society. On a more practical level, it is not entirely unheard of for correctional officers to show their badge or to proffer information about their profession to a police officer who may have stopped them for a traffic infraction. My understanding is that the same courtesy one police officer may extend to another in a similar situation would possibly not apply equally in the case of a correctional officer. The same level of collegiality or "tightness" appears not to apply across the two groups.

One further note: It is entirely possible, indeed probable, that there may exist different subcultures within the subculture. Possible examples of these sub-subcultures might be built around issues of gender, age, or role within the organization.

These differences notwithstanding, however, I think it will become clear that there are sufficient crossovers to consider both police officer and correctional officer subcultures as one for the purposes of our discussion.

The discussion in this chapter and the next may touch a nerve. It has the potential to make you feel defensive. On the other hand, some of you may even feel a sense of relief that someone (stupid enough?) like me is prepared to broach the subject!

For you to keep an open mind about this topic, and to give it a fair hearing, will require considerable courage, and intellectual and emotional maturity. Of course, if after reading chapters 5 and 6, and thinking about and discussing their contents, you wish to discount some or all of the ideas, then that is your right. What matters most is what you think. But be aware, it will not be easy for you to keep an open mind. I suspect, based on discussions of this topic with students, that it may be much easier for you to react defensively.

WHAT DO WE MEAN BY "SUBCULTURE" AND "SUBCULTURAL CONSTRAINTS"?

To help us develop a better understanding of the terms **subculture** and **subcultural constraints**, we will consider three sources of information. First, we will turn to a dictionary definition for help. Second, we will discuss the ideas that some faculty members in college justice programs have about these two terms. Third, we will look briefly at what some researchers have to say about them. In the following chapter, we will consider the views of some justice educators, police and correctional officers, and administrators.

First, let's look at a dictionary definition of subculture. The *Collins English Dictionary* (Third Edition) defines subculture as "a subdivision of a national culture or an enclave within it with a distinct integrated network of behaviour, beliefs, and attitudes." The key idea here is that a subculture is characterized by a code of behaviour governed by a set of beliefs and attitudes.

Faculty Opinions

Second, let's consider what faculty in college justice programs have to offer in helping us understand these terms. It is significant, I think, that all of these faculty members have enjoyed careers in the justice field, and since becoming educators have supervised students extensively on their field practica. This activity has enabled these teachers to keep up with developments in the field. It would be reasonable to assume, therefore, that their opinions continue to be current and informed.

The faculty opinions were expressed in a survey that I conducted. Faculty were asked to respond to the question "What do you mean by the term *subculture*?" Their responses were interesting and informative and will provide stimulating topics of discussion for you and your class colleagues. Keep in mind that in all cases the faculty were defining *subculture* as it relates to the justice field. (In the next chapter we will consider what these justice educators said to me about this topic in one-on-one interviews.)

As you read this chapter, take some time to reflect on your own personal experience. What you think and feel about your workplace and how it impacts upon you as an individual is of critical importance to your development as a human being and as a police or correctional officer.

Here are some samples of survey responses. One teacher wrote, "By subculture, I mean a different, distinct, and separate cultural phenomenon—a culture other than the mainstream culture. One with its own values, beliefs, mores. The major value of this subculture is to 'protect your own butt,' and, secondarily, 'to protect and stand by your co-workers.'"

It is interesting to note what this person claims are the two main values characterizing police and correctional officer subculture. Clearly, these two values more often than not characterize *any* workplace. The difference, however, between, say, a justice workplace and an automotive plant or an insurance office, may be the *degree* to which these norms characterize those particular workplaces. As we shall see later in this chapter, arguments can be made that because of issues around the safety and security of officers in the justice field, the requirement to adhere to subcultural norms is stronger.

"The subculture," claims another faculty member, "consists of a set of beliefs, habits, attitudes, and even schemes which determine in large part what line workers [officers] have to do to survive in their occupational environment." Interestingly, this person added in his response that in such a subculture these beliefs and attitudes are conveyed *informally*. They constitute what is essentially a code of conduct for officers, known by all officers but normally not formally communicated. The code becomes understood, almost by osmosis, and then is reinforced when a non-conforming officer is chastised (ostracised?) by her colleagues. We referred in Chapter 1 to the view of Professor Frederick Desroches that "novices [new police officers] entering the group are confronted…with a set of norms and values to which they are exposed and are expected to conform." There are some clear connections here between this particular faculty member's ideas and those of Desroches. The faculty member talks in terms of what "officers have to do," and Desroches talks about an expectation of new officers to conform to what we sometimes call a "code of conduct." Both these people make it sound as though one doesn't have much choice in the matter. Perhaps, as we noted in Chapter 1, that is why two researchers, Grossi and Berg (1991), talked in terms of correctional officers having to "compromise their personal values and interests." If this kind of thinking has any merit, isn't it curious that an informal code can be so powerful?

We have referred to an unwritten, informal code of conduct governed by a set of expectations. One faculty member, in explaining his views on subculture, talked in terms of "a group of persons who adhere to an unwritten code of conduct which governs certain areas of behaviour. The subculture is also the vehicle through which values and morals are conveyed by their proponents, and the rewards/punishments imposed for conformity/non-conformity."

Many other faculty members responded in a similar vein to the survey question asking them to define what they mean by "subculture." A recurring theme was that a subculture is generally characterized by expectations that the group has regarding the behaviour of its members.

Researcher Opinions

Third, many of the researchers in this area tend to agree with this kind of thinking. Sociologist Theodore Mills (1967), for example, talked about the norms of the group being "ideas in the minds of members (of that group) about what should or should not be done by a specific member under specified circumstances" (p. 74). Three other researchers— Alan Coffey, Edward Eldefonso and Walter Hartinger (1982)—in discussing police subculture, defined it "as a kind of shorthand term for the organized sum of police perspectives, relevant to the police role" (p. 141). Two others, Darwin Cartwright and Alvin Zander (1968), talk about the pressure that groups can bring to bear on their members: "But it is clear that groups can, and often do, apply pressure on their members, so as to bring about a uniformity of beliefs, attitudes, values, and behaviour.... If a cohesive group has developed a standard or a norm, it may exert strong pressures on any member who attempts to deviate" (p. 145).

Perhaps you will agree, if any of the foregoing ideas are worthy of serious consideration, that a discussion of subcultural constraints has a place in a book about ethics. If the group to which we belong, in the interests of maintaining group solidarity, expects us, even occasionally, to give up our right to make our own moral decisions, maybe we should be concerned.

Of course, there are many other aspects of the development of a subculture that need to be taken into consideration, and again the researchers can help us develop an understanding of these.

A SUBCULTURE NEEDS PERCEIVED ENEMIES

Some researchers express the view that for a subculture to become established, the subcultural group has to be anti-something or another. In other words, that there has to be a common "enemy." Members of the group join together in their opposition to the "enemy" and this, in turn, can galvanize the group. For example, sociologist G.L. Webb (1978), in addressing correctional officer subculture, talks about that particular group being anti-inmate, anti-administration, and anti-prison social worker. Criminologists Grossi and Berg (1991) talk about "hostility directed at officers by inmates, their [inmate] families, and often even the public," thereby making them the "enemy."

In the police context, similar sentiments have been expressed by other researchers, including Bittner (1980) and Goldstein (1990). Police officers, for example, when they perceive themselves to be under siege from the public over an alleged act of professional misconduct, often see themselves as being misunderstood. In other words, they feel only a police officer can understand police conduct—"They" (the public) do not understand "us." In the following chapter there is a discussion of one former police officer's description of what he calls the "musk ox syndrome." His depiction graphically portrays how members of a police force may react in the face of public scrutiny or criticism. It doesn't constitute a quantum leap, I think, to see how this kind of dynamic can have the effect of galvanizing a group.

In summary, then, we have briefly examined what we mean by a subculture in the justice field. We have suggested that it is characterized by a set of norms, communicated informally *to* all its members *by* members. These norms are said to govern the conduct of those members if they wish to preserve their membership in the group. We have also discussed the idea of a subcultural group needing—for its formation and continuing existence—to have a real or perceived enemy. We have suggested that this can galvanize the members of the group into a "solid" entity. In the case of police and correctional officers, the "enemy" is often seen to be the criminal element, the officers' respective administrative groups and, perhaps to a lesser degree, the public.

Reflections

Would you agree that the issue of police and correctional officer subculture is a highly sensitive and emotive one? If so, why do you think this is so? What feelings have you been having since you began thinking about the topic?

How would you define *subculture* in the justice context to someone who had no idea what you were talking about?

What are your views on the idea of an unwritten code of conduct governing the behaviour of police and correctional officers at work?

ARE THERE DIFFERENCES AMONG WORKPLACE SUBCULTURES?

You will have noticed that for the purposes of this book the notions of "police officer subculture" and "correctional officer subculture" are used interchangeably as though they are, to a large extent, one and the same thing.

More research has been done on the former, so we know a little more about subculture in the police context. From this research it may be argued that the subculture in policing is stronger than it is in correctional work, and that the influence of the subculture tends to remain relatively constant across differing police authorities. The idea of a police brotherhood is believed by most to extend across police departments. Perhaps that is why in countries such as Canada, the US, Australia, and Britain, to name a few, it is customary in sad instances of a police officer fatality for officers from other forces to attend the funeral to express support and solidarity. It is also common for correctional officers to send delegates to such funerals.

Subculture Strongest in Maximum Security Institutions

With respect to corrections, my research indicates that the strength of the subculture tends to correlate with the security level of a correctional facility. It tends to be strongest in maximum security institutions, and weaker as the security level is reduced. However, even in minimum security institutions, an officer subculture is still evident, although the constraints are generally not as strictly enforced. For example, it is conceivable that officers working in minimum security facilities may see the inmate population not so much as the "enemy" but as a group of individuals who may benefit from their incarceration. Generally, this is most unlikely to be the case in maximum security institutions, and somewhat less likely to be the case in medium security facilities. This view was confirmed in a discussion I had with an institution administrator. He said, "The nature of the subculture and how ingrained it is, in my opinion, depends on whether you are in maximum security, medium security, or minimum security. It is just like the inmate subculture which is highly ingrained in maximum, less so in medium, and less in minimum."

That said, police and correctional officer subcultures share, in my view, a number of common characteristics. These allow us to regard them as sufficiently similar to warrant considering them together. For the sake of simplification, then, and to avoid the risk of becoming bogged down in a debate over the finer differences between police and correctional officer subculture, or over the differences between subculture in a maximum security setting and a less secure institution setting, we will consider them to be essentially similar.

There may be a place in another book for a discussion about the differences, but here we are simply looking at the idea of police and correctional officer subculture in general terms, and how its constraints may influence the individual officer's sense of autonomy in the workplace.

Distinguishing Characteristics of Officer Subculture

What, then, are some of the distinguishing characteristics of police and correctional officer subculture, and why is it sometimes claimed that this subculture is different from that in most other workplaces?

Support for seeing the two subcultures as similar is offered by Grossi and Berg (1991), to whom we referred earlier. They write,

> Those who have studied the internal workings of correctional institutions may recognize that powerful forces within the correctional system have a stronger influence over the behaviour of correctional officers than the administrators of the institution, legislative decrees, or agency policies. These social forces may be understood as related to a "correctional officer subculture," similar to that found in policing (p. 79).

These two writers list what they call "setting-related stressors" for correctional officers that they claim are "remarkably similar to those commonly associated with police officers" (p. 79).

1. The ever-present potential for physical danger.

2. Hostility directed at officers by inmates, their [inmate] families and often even the public.

3. Unreasonable demands and expectations on their role as correctional officers; vacillating political attitudes toward the institutional role of corrections.

4. A tedious and unrewarding work environment.

5. The dependence correctional officers place on one another to effectively work in the institution safely.

6. The reality that one cannot always expect to act the way either one would choose to, or the way the public might expect one to.

(Point 6 is one of the few references I have found to support the idea that police and correctional officer subculture may impair the ability of the individual officer to make independent moral and practical decisions at work.)

Kelsey Kauffman (1988) is a scholar who researches primarily in the corrections context. In her book entitled *Prison Officers and Their World*, based on a four-year study of the Massachusetts prison system, she writes, "Officers also possess a distinct subculture within prisons. Their own beliefs and code of conduct set them apart from administrators, social workers, and, of course, inmates." She "discovered" the beliefs and the norms that constitute the code of conduct by asking officers two questions: 1) What is the worst thing you could do in your own eyes as an officer?, and 2) What is the worst thing you could do in the eyes of other officers? Nine norms were identified:

1. Always go to the aid of an officer in distress.

2. Do not "lug" drugs.

3. Do not "rat."

4. Never make a fellow officer look bad in front of inmates.

5. Always support an officer in a dispute with an inmate.

6. Always support officer sanctions against inmates.

7. Do not be a "white hat." [This denotes any behaviour suggesting sympathy for or identification with inmates.]

8. Maintain officer solidarity against all outside groups.

9. Show positive concern for fellow officers (p. 85).

Would it be reasonable to suggest that if we substituted the word *criminal, offender* or *suspect* for *inmate*, many of these norms would apply equally to policing? Clearly, the predominant theme here is that of *solidarity* of the officers, and of support and protection of one another in the face of the common "enemy." The enemy is the criminal element and, to a lesser extent, outside groups and the public.

Let me offer an illustration of this point about officer solidarity. A newspaper article entitled "Four Officers Suspended in Probe" indicated that four police officers had been suspended pending a probe into allegations that they had planted drugs on suspects (*Toronto Star*, 17 March 1995, p. A6). The article included the comment "Detectives from the force's internal affairs unit launched their probe two weeks ago after a constable—whose identity is a closely guarded secret—complained that a veteran officer from 12 Division's B Platoon had been involved in illegal conduct.... Meanwhile, the name and division of the police officer who made the complaint are being kept secret. Police sources say they don't know what motivated the constable to come forward with the allegations at this time."

It would be inappropriate to read too much into this newspaper account. However, on the face of it we may have an example of one police officer feeling the need to "break ranks" and report behaviour of colleagues that he considered unacceptable or even illegal. That the reporting officer's identity was kept a "closely guarded secret" should come as little or no surprise, because of the potential for his being ostracized by the group. (As an aside, there is a conundrum here for the administrator to whom the officer reported the wrongdoing. The administrator has to weigh keeping the confidence of the reporting officer against the right of his or her accused colleagues to know who the accuser is. We shall return to difficulties like this one later when we discuss the ethical dilemmas). If Kauffman is correct, this reporting officer failed to comply with at least eight of the nine subcultural norms. Please look at the norms and identify for yourself how many you think were breached by the officer.

Let me give you another example. This one involves a case of a police officer who testified that he had falsified notebook entries to bolster a case against a man he and his partner had arrested on drug charges (*Globe and Mail,* 23 October 1996, p. 96). The two officers fabricated evidence against the accused person. The constable's partner suggested to him that "if I mentioned [Mr Brown] had the crack pipe in his mouth, it would work better as grounds for the arrest." The 26-year-old officer pleaded guilty to attempting to obstruct justice and was sentenced to jail. In his defence, the officer said, "I went along with the flow...I went ahead with it...I didn't want to rock the boat." And then he said, "I wouldn't have much of a career left, as far as people working with me, if I had blown the whistle. You're supposed to be a team player."

In the mind of this young officer, there was a strong constraint to remain solid—even in the face of what he perceived to be wrongdoing.

You are correct if you are thinking we haven't yet identified what makes a subculture in the justice context substantively different from that in other workplaces. One writer, Vernon Fox (1983), talks about one difference in his book *Correctional Institutions.* He notes that much has been written about the loyalty of police to their colleagues and affirms that such loyalties are also common in the corporate world. "Nevertheless," he says, "this loyalty appears to be stronger and have more meaning in the criminal justice system than in most other groups" (p. 162). Some of you might disagree with Fox. You might argue that the subculture in General Motors, for example, is similar in many respects. You may well be correct.

One important difference, however, is the anti-criminal element. Most other workplaces see the customers in positive terms, often as "friends." If they did not, they probably would not survive for very long! Offenders, on the other hand—the "clients" of police and correctional officers—excite a different response, normally one characterized by antipathy. And this antipathy can create a bond among police and correctional officers as they see themselves working together against the "bad guys."

One other even more significant factor may account for the difference between subculture in corrections and policing and that in other workplaces. In policing and corrections there is an ever-present potential for danger—life-threatening danger. The potential for danger exists because of who the "clients" are. There are significant issues around personal safety that generally do not characterize most other workplaces. And it is this ingredient of danger that may lend credibility to the view that police and correctional officer subculture is stronger.

If the subculture is stronger, it may be reasonable to think of it as a more compelling constraint on the conduct of its members.

Danger Reinforces the Subculture

Some interesting work was done in the 1960s on group solidarity by sociologist Irving Janis. Janis, basing much of his theorizing on war incidents, supported the idea that the presence of danger tends to increase the solidarity of a group. For example, he wrote, "It has long been known that when people are exposed to external danger they show a remarkable increase in group solidarity. That is, they manifest increased motivation to retain affiliation with a face-to-face group and to avoid actions that deviate from its norms (p. 80).

This, I believe, is the key difference between police and correctional officer subculture, and subculture in most other workplaces. The individual officer's welfare, if not her life, can sometimes rest in the hands of a partner. The willingness of colleagues to respond, support and provide "backup" in an emergency is a basic tenet of the police and correctional officer's role. This, I would suggest, can in turn lead to a bonding that goes beyond the kind one might expect in most other workplaces.

RATTING: THE LOYALTY NORM

Security issues and the way in which individual officers have to rely on each other for their safety makes the requirement to remain loyal to one another the key norm in their subculture. There is a strong prohibition against what is often called "ratting" or "ratting out" a colleague. Coincidental with writing this section of the book, I was conducting a seminar at a secure, young-offender facility. The officers and I engaged in a conversation about this topic. At the end of the day, one of the staff came up to me, and surreptitiously pulled from his pocket his electronic key fob, on one side of which he had placed a picture of a rat. He had reported a colleague for hitting a resident, had been labelled a rat, and had decorated his key fob in this way as a humorous attempt to neutralize the stress of his "label." It is not unknown in corrections or policing for rather severe consequences to be inflicted upon a colleague who has been labelled a "rat." These consequences can range from being "put on the grease" (shunned by colleagues) to threats of personal or property harm being made—or carried out—against the "rat."

The television news program *60 Minutes* carried a story that may be pertinent to this discussion. According to the story[1] (CBS News, 30 March 1997), two officers named Rigg and Caruso at California's Corcoran State Penitentiary blew the whistle on what they considered to be unethical conduct by their colleagues. The two officers claim that inmates from different ethnic groups were set up to fight gladiator-style in the small exercise yard of the Special Handling Unit. It was not unusual, they claimed, for officers to place bets on the outcome of the fights. When the fights got out of hand, they claimed, officers would on occasion shoot at the inmates. Since the institution had opened in 1988, eight inmates had been shot dead by officers and numerous others had been wounded. Inmate Preston Tate, it was claimed, was told after he had hit an officer that he would never get out of the facility alive. On the program, videotape footage was shown of Tate and another inmate of the same ethnic origin fighting two other inmates of a different origin. Tate was shot and killed in the melee even though the videotape clearly shows that he was not the aggressor and that he was away from the fight when he was shot. Lieutenant Rigg, one of the whistle-blowing officers, described the shooting of the inmate as murder.

During the interview with the two officers, both said they had been labelled as "rats," as "no-goods," and both had had their lives threatened by officer colleagues. Even though they had been transferred to other institutions the labels travelled with them, which made their lives in their new workplaces as miserable as they had been at Corcoran. At the con-

---

[1] Since the airing of the *60 Minutes* program, the allegations made by Rigg and Caruso have been substantiated.

clusion of the story, interviewer Mike Wallace, at the new place of work of officer Richard Caruso, hailed down the vehicle doing the institution perimeter check and asked the officer driving the vehicle if he would shake hands with Caruso. It was a telling moment when the driver looked incredulously at Wallace, shook his head in a gesture of defiance, and drove off. Breaking ranks and being labelled a rat can be an unpleasant, unsettling experience.

In the mid-1980s, a female correctional officer at a medium security correctional institution reported some of her colleagues for sleeping on night shift. She had reportedly approached them and said several times that she felt unsafe and vulnerable when they were asleep. She told them that if they did not refrain from doing so, she would have no alternative but to report them to the superintendent. They continued. She reported them. The consequences for her were severe. Graffiti was written about her on some of the walls in the institution; she received harassing phone calls and letters; her car was deliberately damaged in the institution parking lot. The campaign of terror by her colleagues culminated in some bricks being thrown through the windows of her home. It appears she paid a heavy price for doing what she considered to be the right thing. In this case, the right thing for her was to attempt to ensure her own safety and that of the institution.

It was many years ago now, but I remember experiencing a three-week ride-along with a police force. In a ride-along program, as many of you will know, a civilian is taken along in a cruiser to observe police officers at work. I had been invited to do so after spending several weeks conducting in-service training sessions for this particular police force. I had also gotten to know several officers through a criminology course I was teaching at a local community college. The officers had accepted me and became very open with me about their work, about their rites of passage and about what they considered the "perks" of their job. I remember riding along with one particular officer who appeared not to be well-liked or well-respected by his colleagues. I noticed that the other officers were invariably slower to respond to his calls for assistance. I questioned them about this apparent antipathy because I could not make any sense of their reaction to this officer. He struck me as being a professional, principled, family- oriented individual. In retrospect I can see that he was deemed by his colleagues not to fit in. Several years later, I think that I can see why they reacted to him in this way. He was not one of "them." He marched to the beat of a different drummer. He refused to participate in the on-duty activities of some his colleagues, some of which, I remember, stretched the limits of professional conduct.

If we were to select those who should appear in the pantheon of officers who have placed themselves at risk for reporting the misdeeds of others, Frank Serpico would be high on the list. As a member of the New York City Police Department, he placed himself in great peril for exposing the rampant corruption within that agency in the early 1970s and for providing exact details about the corruption before the Knapp Commission. Today, more than a quarter century after that testimony, Frank Serpico continues to evoke considerable emotion in large parts of the police community. Most officers readily describe him as a hero, while many continue to see him as a "rat."

In the late 1990s and early 2000s an interesting case unfolded in Sydney, Australia. A former assistant police commissioner brought a defamation action against former police board members. One of the witnesses who gave evidence was a former police officer Deborah Locke. Mrs Locke said that passing on information about the corruption of some of her colleagues to the internal investigations unit led to her fearing for her life. She said there had been a leak from the investigations unit to her corrupt colleagues and that they were out to get her. Mrs Locke went on to testify that she had a meeting with the former assistant commissioner and at that meeting he made it very clear to her that she was a problem, that she was an *outcast*. She further testified that one of her supervisors said to her, "we don't like this sort of thing, we are a brotherhood."

During the summer of 2002, a 20-year veteran police officer attending an ethics course in the United States raised his hand and asked to speak. When he did, he described a series of events in which he recently found himself embroiled, including his decision to wear a wire (a concealed voice recording device) to help his department make a criminal case against a corrupt police officer. He provided a measured and calm outline of the details of the case but became emotional as he described the way some fellow officers and supervisors reacted upon learning of the assistance he provided to Internal Affairs. "I couldn't believe it," he said. "Even though this cop pleaded guilty to his crimes, I was ostracized...shunned...by a number of other officers and even command personnel. It's been a very painful experience."

This will probably not come as any great surprise to you, but I think it is interesting that the inmate subculture is also characterized by what we will call the "loyalty norm." Anyone who has worked in the justice field, particularly those of you who have worked in correctional institutions, will be familiar with the way in which inmates are quick and sometimes merciless in weeding out from their number those they consider to be "rats." This is one of the main reasons most institutions have protective custody or super-protective custody wings, ranges or cell blocks—so that inmate "rats" can be segregated for their own protection.

Permit me to make a personal observation. In my ethics seminars I discuss the issue of subculture, and the police or correctional officer participants are generally quick to identify the "ratting" norm as a key one. In addition to the term "rat," they use, almost interchangeably, terms like "narc" or "fink." In Australia, the act of reporting a colleague is often called "dobbing in a mate"—a sin in the Australian culture considered almost equivalent to blasphemy! (It would be interesting for someone to investigate whether or not this strong societal norm resides in Australia's history of being settled by convicts.) In the seminars, the participants, when asked about the effects of the "ratting" norm on their individual sense of what may or may not be the right thing to do, become deeply reflective. And this happens almost without exception. It is as though the participants are taking personal stock of where they are as individuals with respect to this issue. When I question them about the apparent change of "chemistry" in the room, they generally confirm that this is precisely what is happening—they are engaging in a personal check regarding how much of their individuality they may have conceded to their colleague group.

A NEED FOR PERSPECTIVE

It would be wrong to discuss this issue of officer subculture and to fail to point out that the subculture has some extremely positive qualities. The mutual support and protection it is likely to engender in crisis situations is essential to the well-being of the officers. The camaraderie often experienced following a crisis situation is sometimes essential to the emotional and psychological health of officers involved. The feeling of belonging to a group whose members understand the peculiar and difficult demands of the profession can be gratifying and comforting, especially when the group or the profession feels under siege from the public.

This camaraderie and support that police officers have with and for each other is graphically expressed in an article entitled "Courtroom 'Tense' for Detective in Shooting" (*Toronto Star*, 7 March 1997, p. A6). A detective, charged after a teen was shot in the head and killed, was supported by "more than 100 off-duty York Region officers."

One justice educator I was speaking with talked about this positive side. "The subculture at times can be a strength. At times, you may be asked as an officer to do some fairly disgusting things…. Once the incident is over, you still have your 'family' to support you." Continuing this thought, another officer said, "The subculture serves you well. There is a lot of camaraderie. I think people do look after each other. There is a recognition coming from one's peers that is more valued than that coming from administrators."

It would be inappropriate to expect individuals involved in the dangerous professions of policing and corrections to "go it alone." Indeed it would be, and is, a distinct worry for both colleagues and managers to have to work with a maverick, "loner" officer.

This chapter and the next are not about breaking the subculture. They simply raise the question as to whether, on occasion, the officer subculture can have a darker side; a side where too much of a person's sense of independence and individuality is given up to the group, even when the group may be acting against the moral suasion of that individual. Remember, in Chapter 1 we called it de-individuation.

SUMMARY

To summarize, in this section of the chapter we have discussed whether there are sufficient similarities between subcultures in policing and corrections to allow us to consider them, for the purposes of this book, essentially similar. It was suggested that an argument could be made for doing so, although the finer differences between them would make for an interesting research topic.

By way of explaining the similarities between the two subcultures, certain norms, common to both, were identified.

The essential similarities are created and reinforced, we suggested, first, by the inherent danger in both roles—there are distinct and ever-present risks to the safety and well-being of the officers; second, by the reality that in these two subcultures the general feeling toward the "client" is one of antipathy, a situation not generally found in other workplaces in government, industry, or commerce; third, by a shared antipathy toward the public "them" who do not understand "us."

Because of these factors, we have suggested that policing and corrections subcultures are, in some significant ways, different from "regular" workplace subcultures, and that the most significant difference—the threat to the officers' safety—ensures to a greater degree than in other workplaces that the loyalty norm is respected by individuals and enforced by the group.

Reflections

What are your views about whether police officer and correctional officer subcultures are generally similar?

What are your views about the idea that there are some significant differences between police and correctional officer subculture, and subcultures in most other workplaces?

What do you think of the nine correctional officer subcultural norms identified by Kelsey Kauffman and the six stressors identified by Elizabeth Grossi and Bruce Berg? Which ones, if any, do you think would apply to police officers?

chapter six

Subculture: What the Practitioners Think

I know what the officer code is and it is very strong. A person can come into this work with all good intentions, but a lot of people just give up and fall into the pit [the subculture].

—Senior federal institution administrator

I would say the peer pressure is very strong. Very few officers would go against that peer pressure.

—Senior institution clinician

[If you were not loyal] you would probably be driven out of the institution very quickly. You would have so much pressure on you that you would resign.

—Correctional officer

It's important to be a member of the brotherhood, whether you're male or not, it's a brotherhood, you subscribe to the rules of the brotherhood...and you do not violate the rules. Time and again we see people being ostracised because they have broken the rules of the group. You had better not break the Code of Silence.

—Police trainer

I was once told by an officer I respect, "We have to be our own person and develop our own methods of work, yet we must manage these ideals within the framework of the institution."

—Justice student after field practicum

CHAPTER OBJECTIVES

After reading this chapter, you should be able to

- Explain why it may be risky for a police or correctional officer to talk openly about the group norms that govern his conduct
- Identify three key ideas that practitioners have about officer subculture and subcultural constraints
- Explain why subcultural constraints may make it difficult for officers to make autonomous moral decisions at work
- Explain the terms **loyalty** and **solidarity** in the context of a justice subculture
- Explain why there may be a link between officer and offender subculture
- Identify one reason why some officers may be able to take moral stands on certain issues and not be ostracized by their colleagues
- Explain why members of a subculture need to identify common "enemies," and what effect this has on the group

A SENSITIVE TOPIC

Much of the material for this chapter has been gathered in one-on-one personal interviews. The interviews were conducted with justice educators, police and correctional officers, and correctional institution and police administrators. The transcripts of many of the interviews constitute a part of the qualitative data gathered for my doctoral dissertation. The material is included here with the written consent of the respondents. Great care has been taken not to identify any police force, correctional jurisdiction or institution, or individual officer. The willingness of these people to speak to me is much appreciated. They have contributed significantly to my learning, and I hope their insights will contribute to your learning also.

The issue of subculture in policing and corrections is a highly sensitive and emotive one. In their willingness to express their views on this topic, some of these individuals, particularly officers, took extraordinary risks. The nature of these risks may become more apparent as the chapter unfolds.

To illustrate these risks, let me recount part of a conversation I had with one officer. As we were concluding the interview I asked the officer if he had any last thoughts about our discussion. The officer confided, "It is difficult. I have found this a difficult interview. I'm finding it [the subculture] is something that you don't talk with everybody about, and it is just, I don't know, it's…." "It's a question of trust," I finished, "and you don't know me from a hole in the ground." "That's it right there," the officer replied. "I think, too, because I am at work I have this shield up around me to protect me…. The co-workers are the only ones you can really trust."

Another officer, in response to my invitation to make some final comments on our discussion, said, "This had better not get out or I'm in deep doo-doo."

It should be pointed out that the officers interviewed were selected randomly; it was not the case, by way of allaying any suspicion you may have about the process, that these officers were handpicked in some way. When you read some of their comments, you may be tempted to think they were selected for interview because they had been identified as loose cannons or had already been targets for ostracism by their colleagues. This was not the case, either. Before interviewing them, I had little or no idea of what their perspectives were going to be, or what risks they would take in sharing their ideas with me.

This chapter is organized around the various points of discussion in the preceding chapter. We continue to explore the terms *subculture* and *subcultural constraints*. We discuss in greater depth whether police and correctional officer subculture is different from subculture in other workplaces.

We continue to explore the links between officer subculture and criminal subculture. We also look at the idea that the subculture is stronger in the justice context and that, as a result, the norms are more compelling on individual officers.

It should, of course, be kept in mind that this discussion does not constitute the definitive word on any point discussed. It is included here simply as a means of fleshing out the discussion in Chapter 5. By including and thinking about some of the ideas of police and correctional practitioners, we may be able to develop a broader understanding of the issue. In this chapter, their perspectives have been interlaced with researchers' ideas and with some case studies, by way of providing a context for these perspectives. The discussion will, I hope, serve to provide further background for your thinking and debating.

SUBCULTURE AND SUBCULTURAL CONSTRAINTS

Throughout the interviews with the educator, officer and administrator groups, certain terms kept recurring:

bonding	unwritten code	put on the grease
adherence	conformity	ratting
code of silence	sticking together	family
code of behaviour	non-conformity	social control
officer code	loss of individuality	ostracized
what officers have to do	loyalty	peer pressure

They are, I would suggest, key to our understanding of subculture in the justice context. We should acknowledge, of course, that they can also be key to subcultures in other workplaces. What we have to keep in mind is the *degree* to which these terms characterize a particular subculture. As I have suggested—and this is borne out by the discussion that follows—they apply to a high degree in police and correctional officer subculture, especially when we consider maximum security institutions. They would also apply, to a lesser degree, in medium and minimum security facilities.

What did the individuals I interviewed have to say in general about subculture and subcultural constraints?

The officer subculture was defined in various ways. One respondent, for example, talked about the subculture as an unwritten but organized set of values that is communicated among the staff. Others talked about the code as an informal one, but one which, nonetheless, is known by all officers. Despite its informality, and despite the fact that it is unwritten, the code is often thought of as having a strong influence on the behaviour of officers. One experienced administrator, who had himself come up through the ranks, described the subculture as a force that has the potential to create a "major impact on the judgment of officers."

The Importance of Remaining Solid

The **loyalty** or **solidarity** issue seemed to loom large in many of the responses. One justice educator, in describing subculture in the corrections context, said,

> It seems to me that what you get is that same "protect your own self and your partners" mentality that I see in other areas of law enforcement…where regardless of how serious the violation [is] of either a code of ethics or a violation of policy, people have convinced themselves it is wrong to "rat."

Another educator talked in terms of there existing a very strong expectation to remain "solid." Some people talked about the loyalty or solidarity requirement as though it were part of the unwritten code. "I believe," said one person, "there's a code among officers—you don't put it on a fellow officer. If you go outside the norm and inform, I think you'll have a great deal of difficulty." Another said, "The downside is that there is a tremendous amount of pressure to conform with one's peers in ways that could put one at great, I'd say ethical, and sometimes even in legal, jeopardy."

What is this pressure, and what might the consequences be of resisting it?

One educator, a former officer, said that "if you choose as an officer not to conform, you would be brought under a significant amount of social control within a very, very brief period of time. I have situations where people have had to deal with crisis situations on their own because they are now perceived as a loner, and as somebody the rest of the officers don't want to have anything to do with, because they are not conforming."

The officers invariably confirmed this idea that there would be considerable risk of social isolation for a colleague labelled as a "rat." "If you rat on somebody, you tend to lose a lot of friends, and, basically, you are left with no one to turn to for anything," said one. Another officer said, "You wouldn't want to 'rat out' on a fellow officer. It is something you just wouldn't do. You would be labelled a 'rat.'" I asked another person what the consequences might be, if any, for an officer reporting a serious infraction by a colleague. "I believe I would be shunned. I would be given a hard time. I would be an outcast. I would be labelled a rat." This shunning, as this officer put it, was often referred to in the interviews as "being put on the grease." For example, one officer said, "You put them on the grease. You don't talk to them, you don't sit with them. You treat them like outcasts."

The rule of thumb—the basic expectation—said another officer, was that "you keep your mouth shut and do not rat other people out." He indicated that the rule was that "if you saw a colleague doing something wrong, that was his/her business." As a brief aside here, one of the more disturbing aspects of working with justice students prior to, and after completion of, their field practicum is that one sees a value shift in individual students from an

"I don't think I could let that go" (about a serious officer indiscretion) attitude prior to placement to a "so long as I don't do it—what she does is her business" attitude afterwards.

A trainee police officer recounted to me some advice given to him during his first few days at a police detachment: "The first day I was there, you know, after having met me and that type of thing, they [the officers] took me in their office and basically had a chat with me. They said whatever you see here, you keep your mouth shut, you don't say nothing [sic]. I kind of looked at that two ways. You know, you see a lot of stuff the general public shouldn't be told about, but also maybe there was a kind of double meaning there. Maybe they were also telling me, whatever you and the other cops are up to out there, don't be going and squealing to his boss or something."

One correctional officer in a maximum security institution went so far as to say that it would be virtually impossible for someone declared to be a rat to survive in that facility. He also indicated it was equally unlikely that such an officer would survive in another institution because the officer grapevine, much like the inmate grapevine, would send word to that institution that this staff member was not to be trusted. Of an officer declared to be a rat, he said, "There is no way you can last more than a month or two months with the type of abuse you would take in this institution." He said the abuse would probably take the form of the "rat" being treated with "dead silence," having letters mailed to him with "rat" on the address, having her tires slashed, or "whatever." When asked to elaborate on what "whatever" may have signified, the officer declined to do so.

Report Writing

To draw this section of the chapter to a close, another officer was direct about what expectations the officers have of each other. "Say a situation happened and something didn't go down the way it was supposed to, and an officer is in the wrong. You don't bring this to anyone's attention, you stick together as a unit and you write reports as a unit." An administrator said something similar: "It is not unusual to have people sit down collectively and write reports. The individual who deviates away from his peers runs the risk of ostracization. They run the risk of outright intimidation from time to time."

What can also happen—and this is a fairly common theme in the interviews—is that if you see a wrongdoing by a colleague and you are questioned about it, you will simply respond that you did not see anything. What is implied here is that there may be times when individual officers are required to write reports or respond in a way they know to be patently false. They know that what they are writing is essentially a lie and they may feel bad about it, but the expectation is that they become part of an event that is unethical or illegal—or face the consequences.

These are difficult circumstances under which to make personal, individual moral decisions.

I don't say this by way of judging you or your colleagues. I feel for you as you wrestle with the fine tension between the basic human (and professional) need you have to be accepted by your colleague group, and the need to be able to do what you consider to be the right thing in a given situation. I know your struggle. I have been there. I don't set myself up here as a paragon of virtue, or as someone having a monopoly on wisdom. I raise the issue of subcultural constraints and individual moral decision-making because I think it should be raised, as difficult as it is to do so. But I raise the issue in a spirit of compassion, not of judgment.

It is important at this juncture that we add some riders. It would be wrong for anyone to suggest that there are no individual officers who have managed to achieve a status with their colleagues such that they can take their own moral stands and survive in their groups. Generally, these are officers who are experienced and respected as professionals; they are people who have proven themselves in dangerous situations and are seen as colleagues who can be counted on in a crisis. On occasion these individuals can, and do, take risks with their group that younger officers dare not take.

It would also be wrong not to add that there are sometimes significant individual differences even among the younger officers. Some are stronger than others, and are able to take more courageous stands, despite the subcultural pressures against doing so.

To conclude this section, I would like to note a comment made by one of the reviewers of the manuscript for the second edition of US version of this book. He was a police officer and is now a police educator. He said: "The author's description of the loyalty and solidarity among police officers is very accurate. This is especially true during times of crisis or stress (officer involved in shootings, high speed pursuits, etc.)."

What do you think? Would it be less difficult for a *former* officer to admit to this phenomenon than someone who is currently on the job?

A Possible Gender Difference

Another interesting idea that came out of my research—and this has been confirmed in several conversations since—is that there may be a gender difference with respect to the personal stands that officers are prepared to take. I hear that women are more likely to take stands. I also hear that this may be due to the fact that they are not subject to the potential for "parking lot justice" that males may sometimes expect. It would be quite inappropriate, of course, to place too much stock in this idea at this time. And we should beware of making gross generalizations based on some personal observations. The idea of gender difference and moral courage may, however, be an interesting topic for further exploration in your classes.

In the meantime, here's some food for thought on this issue. In his autobiography *Moab Is My Washpot*, Stephen Fry, the eccentric British comedian, actor, and author, refers to a critique of a book called *Abinger Harvest* by critic Desmond MacCarthy. In talking about E. M. Forster, the famous author, MacCarthy writes,

> Now, the essentially masculine way of taking life is to handle it departmentally. A man says to himself: there is my home and private life of personal relations; there is my business, my work; there is my life as a citizen. In each department he has principles according to which situations can be handled as they arise. But in each department these are different. His art of life is to *disconnect*; it simplifies problems... The feminine impulse, on the other hand, whether on account of women's education or her fundamental nature, is to see life more as a continuum (p. 138).

Is this why male officers often seem to "check in" their street morality at the front door of the institution or police division when they go to work? Many male officers have said to me that this is how they learn to cope with the stress of their work. They compartmentalize their morality and operate by a different set of rules while on the job. According to MacCarthy, men are acting true to their nature when they do this. Women, on the other hand, are less likely to operate in this way. Because their approach to life, including their morality, is viewed more as a continuum, it is perhaps not surprising that they are more likely to set limits; such behaviour is simply an extension of who they are by nature. And this means that the rules of decency and propriety they live with are more likely to be part of their professional performance.

Some interesting work has been done on gender differences and ethics among police officers in Illinois (Statistical Analysis Center, Illinois Criminal Justice Information Authority, 1994). It would be inappropriate to rush to sweeping conclusions, but the research does seem to indicate that female officers carry to their work as officers a less compromised ethic. For example, in response to the question *Over the past 12 months have you personally observed a fellow officer do the following: Use excessive force? Cover up? Fail to report excessive force?,* a larger percentage of females reported in the affirmative. This was also the case with respect to another question asked, *Over the past 12 months have you personally observed a fellow officer do the following: Stop/frisk to harass? Give false testimony? Write a false arrest report?* A greater percentage of female officers than male officers reported that they had experienced these wrongdoings. In other words, it would seem the female officers were more willing to be honest about their experience. They tend to identify problems more honestly than do their male colleagues.

Dr Gary Sykes, Director of the Center for Law Enforcement Ethics, surmises that the reason for this higher level of sensitivity to issues of right and wrong might be what he calls "feminine morality"; that is, women in all societies seem to have a greater sensitivity to issues of care. As a result, they appear to be committed more to a "rehabilitative ethic" than a "just deserts ethic"—to what Sykes calls a "peacemaking justice." If this is indeed the case, we can see how women's tolerance for wrongdoing may be lower than that of men's.

Of course, a word of caution is important here. We must be careful not to jump to conclusions. There are many male officers who operate in an exemplary way at work, and there are some women who do not. This discussion is offered as food for thought. The concept that men are more able and willing to compartmentalize their lives and their morality is, I think, intriguing. However—and I speak as a man here—it is difficult for us males to face issues such as this one dispassionately; there may be a real temptation to react emotionally.

This topic has been explored in Mexico City. In a city noted for traffic, pollution, and police corruption, the chief of police has taken the highly controversial step of stripping his male officers of the power to write parking and traffic tickets, giving that duty, instead, to teams of female officers. In his view, the only way to break away from the historical and deeply entrenched system that permitted male officers to augment their monthly salaries by accepting bribes is to replace them with women. It will be interesting to see how this experiment plays out, for even some women's rights advocates in Mexico have pointed out that the only difference between male and female officers is that, traditionally, females have not had ready access to power like their male counterparts.

This topic is one that has excited a great deal of interest among readers of the first edition of the book, particularly female officers. Not surprisingly perhaps, they agree with the notion that women tend to be more aware of ethical boundaries when conducting their duties.

For all the anecdotal support for this idea, before we can speak with any real certainty on this topic there is a need for more research. Perhaps a reader like yourself might be inspired to contribute to our thinking on this very important subject. We have to keep the dialogue open on this front because as increasing numbers of women become police and correctional officers we may find the culture beginning to change, too.

Backup

One final note. In some of the conversations I have had with officers and administrators, reference has been made to the idea that an ostracized officer could not be sure of backup in an emergency. It needs to be pointed out—and this, too, has been confirmed for me in less formal conversations with various individuals—that even someone who has been declared a "rat" will almost invariably be afforded colleague support in a crisis situation.

Reflections

What do you think of the comments of your fellow officers included in this section of the chapter? If you agree with all or some of them, would you be prepared to express them to

a. a researcher you did not know?

b. your colleagues?

c. your manager?

d. your teacher?

A number of key terms relating to officer subculture are listed. Which ones, if any, would you agree are significant in developing a better understanding of the norms governing your colleague group?

Why is it, do you think, that an unwritten, informal code of conduct can potentially have such a powerful influence over the decision-making and behaviour of individual officers?

One officer said, "There is a tremendous amount of pressure to conform with one's peers in ways that could put one at great ethical, and sometimes even legal, jeopardy." How would you respond to this statement?

Should there be a place for discussion of officer subculture in your training or education? If so, why? If not, why not?

In your view or experience, are female officers more or less likely than males to take a stand on moral issues that relate to corrections or policing?

What is your view of the idea that men are more likely than women to compartmentalize their morality?

Let us now look at what some of your colleagues and managers have to say about the potential links between the subculture of officers and that of their "clients."

LINKS BETWEEN OFFICER SUBCULTURE AND CRIMINAL SUBCULTURE

This section applies principally to correctional officers, but there are issues that may apply to police officers. In the case of correctional officers, of course, we are talking about imprisoned offenders, and in the case of police officers we would be thinking more of the "street" criminal subculture.

Shared Values

It's a troubling idea, I suppose, that there may be some similarities between the subcultural norms that govern the conduct of those in the criminal world and those that can sometimes govern the conduct of officers. But the links appear to be there, nonetheless. This was a common theme in the interviews I conducted. Consider how Vernon Fox (1983) in *Correctional Institutions* describes prisoner subculture, and think about whether there may be some parallels with the way police and correctional officers govern each other's conduct at work:

> The inmate code in a maximum security prison is made up of the customs and folkways by which the residents protect themselves from the repressive measures of the administration; the inmates have nothing to do with the administration or correctional officers, and they do not "squeal" or "snitch" on fellow inmates. This inmate code, which is fundamental to prison organization, manifests itself in a type of informal control.... (p. 101).

Perhaps it is not surprising, then, that one correctional officer I interviewed said, in his view, the officer subculture is defined by the subculture of what he called the "clients we serve." The officer said, "So a lot of the subculture that you would see among correctional officers is very defined by the inmate subculture." He added, "One doesn't have to work long in institutions to find that people take on the manners of the inmates and their culture."

You will know, whether you are a police officer or a correctional officer, or a student or trainee preparing for either one of these professions, that there is a strong code of conduct that governs the behaviour of the criminal element on the street, as it does the conduct of inmates in institutions. As we noted earlier, any criminal person not considered "solid"—where "solid" means, in essence, fully committed to the group and its code of conduct, including a code of silence—lives dangerously. In an institution, an individual identified as not "solid" often needs to be separated for his own protection. On the street, police informers live equally dangerously. I think it is clear that this value of absolute loyalty to the group can and often does govern officer conduct.

Which Came First—The Chicken or the Egg?

Where this de-individuating value has its roots is a moot point. I call it a de-individuating value because it requires the individual to sacrifice individuality in the interests of the collective. Did this value start with the officers, such that the criminal element adopted the value for itself for its own protection, or was it the other way around? One person I interviewed saw it this way: "I think there is a correctional officer subculture that originates or takes its origins from the inmate subculture. I think it is just as devious and I think it has as much of a dark side as the inmate subculture." Another person said, "So I think it is a learned form of behaviour [the requirement not to 'rat'] that isn't attributed in the system to the inmates, but in my estimation that is where the staff pick it up. They start to play a similar role in how they interact with each other to the way the inmates do."

SOLIDARITY

Whatever the origin of the loyalty norm, one thing seems clear. If you have that sense of solidarity on one side of the fence, you are probably going to have it on the other side. It is, after all, a very human way to react. It probably has something to do with presenting a united, mutually protective front against a common "enemy." One correctional officer expressed this idea well when he said, "I think the inmate code influences us in the fact that you have two distinct groups. You have got the inmates and they will do whatever they can to get by you, and they will stick together no matter what. And we have the same idea; if we are going to have any luck in controlling this place we are going to have to stick together too."

If what we have discussed so far is at least part of the dynamic constituting criminal and officer subcultures, then we can see how the breaking of ranks by a member of either the officer or offender group is not going to elicit a positive response from that member's group.

So what did the practitioners who were interviewed have to say about all this? Not surprisingly, they offered some interesting and, I would have to say, sobering ideas. We don't like to think of ourselves as sharing *any* values with offenders, do we? But in fact we probably do.

By way of introducing their ideas, Mark Baker (1985), in his book *Cops: Their Lives in Their Own Words,* quotes a police officer as saying: "The mentality [of police officers] is an organized crime mentality. It is to be respected in some ways, but it has to be respected from the perspective of criminality, not from legitimate law enforcement. The culture that we come from and the kind of country that we live in, we do respect that kind of camaraderie. But it's magnified in police work. It can go too far" (p. 175).

May I suggest that the police officer, even though he does not elaborate on the point, is suggesting by the phrase "It can go too far" that carrying the norms of loyalty and solidarity in police work too far can, and sometimes does, result in unethical conduct on the part of some individual officers. This sometimes blind allegiance to the loyalty norm can create fertile ground for officer behaviour ranging from poor judgment at one end of the spectrum to corruption at the other end.

In March 1999, some interesting developments unfolded in New York City. Four New York City police officers were indicted for an act of brutality. A Black man, a recent immigrant to the United States, was shot no fewer than 41 times by these officers in the foyer of his apartment building. It appears he had been doing nothing wrong.

The officers came to trial, it seems, only because of demonstrations mounted by concerned citizens in the city—many, but not all of them, Black. Several "big names" participated in the demonstrations, which added to the publicity. Mayor Giuliani had been engaged for the previous few years in a "clean-up New York" campaign; he and his administration seem to have been prepared to tolerate some corruption and brutality by the police force in return for a lowering of the city's crime. Before the demonstrations took place, he did not appear to be overly concerned about the shooting.

Lynne Samuels, a New York City broadcaster, was interviewed by Michael Enright on *This Morning*, a program aired in Canada by the CBC. The interview took place during the week the indictments were due to be heard. Samuels, a White New Yorker, made the point that this clean-up campaign was being conducted at a significant cost—paid mainly by New York City's Black population, who were systematically being subjected to rights and freedoms discrimination by the police.

Then Samuels made an interesting point. She said that the injustices committed by the police were not being exposed because of the way police officers were governed by the requirement to remain solid. She said, "Let's say that only about 5 percent of the 40 000 New York City police officers are bad—that's 2 000 of them running around our city every day doing their thing. The problem is that none of the 38 000 other officers will say anything."

Michael Enright added, "They probably remember Frank Serpico from the 70s and what happened to him when he broke ranks." Serpico, you may remember, was a New York City police officer who attempted to expose corruption on the force. He paid a terrible price for his attempt. He ultimately got shot in the face, and is convinced that he was set up by his colleagues. The solidarity norm would appear to be a compelling one indeed, at least on the New York City police force—if not elsewhere.

A graphic example of this blind allegiance took place in Toronto. Two Metropolitan Toronto police officers were charged with planting drugs on a suspect whom they subsequently charged with possession. In their trial, evidence was given by a young, 26-year-old police officer, who had been convicted (and dismissed from the force) earlier for attempting to obstruct justice in the case by falsifying his notebook entries.

In a newspaper account of the court proceedings, the young officer offered the following in his defence: "I went along with the flow. I went ahead with it.... I didn't want to rock the boat." He said he was very uncomfortable with his actions but continued in this way to justify them: "I wouldn't have much of a career left as far as people working with me.... You're supposed to be a team player" (*Globe and Mail*, 23 October 1996, p. A9).

Maureen Prinsloo, chair of the Metropolitan Toronto Police Board, made a decision to tour some police stations after the two accused officers in the above case were found not guilty of planting drugs on the suspect. In a newspaper account of her visits, Prinsloo is quoted as saying, "[T]he vast majority of officers simply did their jobs." The article continues, "However, she said, when she asked 'Why do you turn a blind eye to this small group who will pull your reputation down?' I got no answer." "To Serve and Protect [the force's motto]," she said, "that is to serve and protect the public. It does not mean protecting the bad ones in your midst" (*Toronto Star*, 21 November 1996, p. A6).

In 1999, a 19-year veteran police officer on the Metropolitan Toronto police force was charged with a series of drug trafficking offences involving high school children. This is an officer who, it turns out, had been in trouble with the police brass on many occasions before. His penchant had been to "get his man" using unorthodox methods—many of them illegal. He was seen by his colleagues as slippery and untrustworthy, but it wasn't until he got involved in the drug dealing that these former colleagues were prepared to say anything.

Christie Blatchford, who often wrote on justice matters for the *National Post* (she now writes for the *Globe and Mail*), a Canadian national newspaper, said this about the situation: "only when the drug trafficking allegations came to light did many of his former colleagues consider that he'd crossed the line, which provides an interesting glimpse into the subculture of even one of Canada's most progressive police forces at the dawn of a new century: Roughing up prisoners may be okay; framing suspects is arguably tolerable if they're believed to be guilty anyway, but dealing heroin to young people? Well, that's beyond the pale" (9 February 1999, p. A1).

Here's the case of a police officer who had coasted along for years, even though many of his colleagues did not trust him; they knew full well that he was given to, let's say, "unorthodox" (read *illegal*) policing methods. But it wasn't until he had crossed *way* over the line, and, indeed, was no longer one of them, that they were prepared to break ranks, breach their code of silence, and overcome their commitment to remain solid. Indeed, you're correct if you are thinking that this does not amount to breaking rank, because the officer was no longer one of them: he had gone "bad"; he was a rogue officer.

(As an aside here, one of my research findings points to the idea that individual officers have differing "lines of tolerance" for colleague wrongdoing. However, one commonly held idea seems to be that officers fraternizing with the criminal element, or willfully engaging in lawlessness, generally will be held accountable.)

One haunting thought for me in the case is this: What if this officer had been held accountable by his colleagues earlier in his career? Might he have avoided sinking to such a painful low? I remember a few years ago reading a letter from the wife of a police officer in an Ann Landers column (no smart remarks, please!).

The police officer had just been fired. He had done something terrible as a direct result of his alcoholism, and his career had crashed to a halt. His wife wrote in her letter something to this effect: *My husband's worst enemies were all those colleagues who covered for him. Why didn't any of them hold him accountable?*

It's a sobering thought, isn't it?

One justice educator, an experienced correctional practitioner, indicated what he considers to be the cardinal norm in both officer and criminal subcultures. He said, "An inmate doesn't inform on an inmate, and an officer doesn't inform on an officer." And another officer said, "Just like the inmates have an unwritten rule that they don't rat out on each other, so it's an unwritten rule that officers don't rat out on each other.... Our code and theirs is basically the same code."

Another educator, a former officer and administrator, took this theme further. "But I also think of the unethical practices with respect to the subculture. I think the prisoner subculture has rules like 'You don't rat'.... Well, a similar process happens for correctional officers where they will be under pressure not to say anything if one of their co-workers does something that is considered to be against the rules. For example, hits an inmate, beats an inmate up, steals from an inmate, misrepresents the facts in a report.... New staff may be pressured to learn very quickly an opposite way of doing things from the way in which they have been taught. You join the system. But you get that way, you pick it up—in the same way police officers do."

It's pertinent that a connection is being made between the subcultural pressures one can feel as an officer and what this person calls "unethical practices." It is critically important that we refrain from making any overgeneralized or carte blanche statements about this connection, but it is a question that deserves serious and open-minded attention.

MORE SHARED VALUES

Some of the officers interviewed talked about other shared values between the officer and inmate subcultures. One institution administrator, for example, talked about how both the inmates and the officers see the sex offender as hanging onto the lowest rung of the ladder in the inmate hierarchy.

Others talked about the way language is shared by the two subcultures. One officer said of his colleagues, "I think they pick it [the language] up from the inmate subculture. It is a strange thing but they pick up the inmate jargon. They talk about 'Don't crack to me,' an inmate expression which means don't talk to me. Both groups use expressions like 'put on the dummy,' 'cheese eaters,' 'laying track,' that kind of thing." Those who have worked in the justice system will also be aware of expressions like "bit" (sentence), "diddlers" (child molesters), "deuce less" (two years less a day), "six" (officer coming), "the hole" (segregation), or "queen" (effeminate male inmate who may or may not be in the process of a sex change), to name just a few terms often used interchangeably by prisoners and officers. Many of these terms, of course, apply principally in the context of correctional institutions.

Police officers also share a common language with the criminal subculture. They, like correctional officers, would use terms like "rat," "fink" or "narc," and "diddlers." In the United States, police officers talk about "stand-up guys" (officers who will protect a colleague who has done something wrong), and "hand-up guys" (officers who will do the opposite). The terms refer, essentially, to officers who are considered solid or not solid. Police officers may also use terms such as "screw" (correctional officer), "joint" (jail), "hopped up" (under the influence of drugs), "roach" (the butt of a marijuana cigarette), and "score" (to purchase drugs). Mark Baker (1985) in *Cops* quotes one police officer as telling him, "So many police terms are picked up from the street world and become incorporated in police mentality [and language]" (p. 199).

Reflections

Who influences whom in the use of this common argot? I put this question to an institution administrator, who stated clearly his view that it was the inmate language that influenced the language of the officers. He said, "You don't have inmates going around talking like correctional officers or acting like them in any way, shape or form."

Would you agree there may be some links between officer subculture and criminal subculture?

If you agree, what would some of these links be? Can you think of some of your own?

Do you agree that officers and the criminal group can come to share a jargon?

SUBCULTURE AND ENEMIES

We have suggested that subcultures are stronger and may require for their existence an identifiable common "enemy." In the case of police and correctional officer subculture we have suggested that the enemy may take three forms. The primary target for officers' antipathy is the offender group, followed in order by the administration, and the public. This is summed up nicely by Mark Baker (1985) in *Cops*, when he talks about police officers' attitudes toward these groups:

> Many cops begin to see themselves as under siege from all sides. They develop a classic "Us versus Them" mentality.... The real reason most police officers socialize exclusively with other

police officers is that they just don't trust the people they police—which is everybody who is not a cop.... Older officers teach younger ones that it is best to avoid civilians.... The "Us against Them" attitude also applies within the ranks of the police themselves. Superior officers from sergeants on up are regarded by patrol officers as "Them" (pp. 175–176).

Anti-criminal

It should be acknowledged that there are two scholars who argue, in the context of corrections, that correctional officers are not as anti-inmate in their disposition as they themselves, and others, sometimes think they are. John Klofas and Hans Toch (1982) developed the concept of "pluralistic ignorance" among correctional officers. Based on a survey of correctional officers, they are of the opinion that there are what they call "progressive" officers (officers who are not anti-inmate) who think they are in the minority, but are, in fact, in the majority. Likewise, officers with non-progressive (anti-inmate) views, who feel themselves to be in the majority, are, in fact, in the minority.

The results and conclusions of the survey are curious. My research and experience present a different picture—that officers in the main are, or become, anti-inmate in disposition. One or two experiences of being "conned" or manipulated by an inmate can quickly change the positive officer into a suspicious one who sees the inmate group as the enemy. And generally speaking, any officer who continues to be essentially pro-inmate in the face of such experiences is often viewed with great suspicion by his officer colleagues.

One administrator I interviewed said, "I know what the code is and it is very strong. A person [new officer] can come into this work with all good intentions, but a lot of people just give up and fall into the anti-inmate routine because it is easier—make your money, go home, and forget about it." And another officer talked about how an officer's attitude toward inmates can change: "It's a funny thing. When you first start here, you tend to sympathize and empathize with the inmates, but after dealing with them on a daily basis, when you are treated badly, in a way you are almost happy if something does happen to them."

One officer made a stronger statement when he talked about officers having "an inherent dislike for inmates, not a feeling of revulsion necessarily, but certainly a dislike for most of the inmate population just simply because they are inmates."

The case of police officers is an even clearer cut one. There is a strong antipathy toward those who break the law, and who represent, potentially, personal danger for the officer. In a chapter of *Cops* entitled "Blood Brothers," Baker (1985) talks about how this antipathy toward lawbreakers and the danger they represent can galvanize police officers into a solid group. "Nobody understands but another cop," one officer is quoted as saying.

Let's take a brief look at what people in my interviews had to say about these antipathies.

One clinical psychologist who works in the justice field talked about how the relationship between officer and offender is characterized by an "us" against "them" attitude, and he attributed the binding together of both officer and inmate groups to this way of thinking and feeling. An experienced administrator and former officer had this to say about institution subculture: "It is 'them' against 'us,' 'them' being the inmates or the administration, depending on the issue." An officer expressed some strong anti-inmate sentiments: "You are dealing with people who go against every principle you have in your life. You are on the other side.... You have got them treating you like you are a lowlife."

Another officer was even more strident in his account: "The minute you walk through the door, you are in a whole different environment.... If the inmates can get something off you, they will lie through their teeth to get it.... You get burned a couple of times by inmates and, yeah, you are going to come out with a very cynical attitude—and that is the attitude you have in this place." One female officer, equally strident, said, "The general public has no idea what kind of animals society has locked up, and you are dealing with these people on a daily basis.... You are no longer part of a polite society, now you are in amongst animals...."

Police officers express similar sentiments about their professional lives and the kind of people they have to encounter during the course of their duties.

Anti-administration

The anti-administration feelings police and correctional officers can develop is illustrated by an event that took place in the Metropolitan Toronto police force in January 1995. A one-day illegal strike of police officers took place in 51 Division. The deputy police chief had decided to order a public inquiry into the conduct of two officers from that division, and this decision precipitated the strike. One of the columnists in the *Toronto Star* wrote this about the incident: "When you try to penetrate the miasma of cop culture, this is what happens. When you cross the line of blind allegiance to cop fraternity, this is what occurs. Politicians are given an ultimatum. The public is threatened with the withdrawal of vital services. And the full ferocity of cop wrath descends on the head of the deputy chief who was already seen—because of his commitment to accountability and procedural reform—as the enemy" (30 January 1995, p. A6).

A correctional officer summed up this anti-administration idea succinctly when she said, "Because the management is after us, you cannot have everybody running around totally against each other." And another officer resorted to the use of inmate jargon to explain where he stood on this issue: "You don't talk to them [the management], you don't 'crack' to them, only in the line of business. You just say the minimal amount that you have to say to them to get the job done."

Anti-public

Finally, here are two sentiments, one expressed by a corrections educator, the other by a police educator who had worked as a police officer for many years. The former talked about what he called the "hunker down" mentality of officers. To explain what he meant by hunker down, he talked about "inmate against guard, guard against inmate, guard against management, guard against the collective community." He added, "Where they, correctional officers and police officers, feel they are misunderstood and that their work is too difficult, and too dangerous, and 'because no one loves them,' what they do is band together and they hide themselves behind the culture of not talking about it, because most of us [the public] don't understand."

The police educator talked in similar terms about police officer subculture. He talked about what he calls the "musk ox syndrome": "You know about musk ox. They are large northern animals in the buffalo family. It's interesting, when they perceive themselves to be threatened, they form a circle with the cows and the calves in the centre of the ring and all the bulls stand shoulder to shoulder with their massive heads out. And, oftentimes what happens is that the subculture within the police is the tendency to pull into a ring similar to that."

The former police officer continued to offer an explanation for this musk ox syndrome. The attitude of police officers, he said, is best characterized in this way: "The only people we can really trust, the only people we are really comfortable with is [sic] ourselves—other members of the subculture. So therefore I should pull into the subculture and don't blame me if I don't talk to you." In expanding further on police attitudes toward the public, he added, "We don't want you to like us, we don't want you to know, or we don't want to know that you like us because it doesn't allow us to pull back into the circle so readily, so the more I can push you farther away, the more comfortable I feel with being part of the subculture."

This anti-public sentiment is best summed up by something an experienced correctional officer recounted to me. "This job can really screw up your thinking about things and about people," he said. "When you are first in the job and you see someone drive into your driveway at home, you say or think to yourself, *Who's that in my driveway?* After a few years, you see someone drive into the driveway and you say, *Who the hell is coming into my driveway?* After several years in the job, you see a person coming up your driveway and you say or think, *Who's that fucking asshole coming up my driveway?*" Those officers whose view of the public deteriorates in this way don't like what they see themselves becoming, but seem to be powerless to resist developing this jaundiced view of folk other than their colleagues.

Anti-public sentiment is confirmed by police behaviour researcher Jerome Skolnick (1975). He suggests that the authority police have over the public, and the danger they feel themselves to be in, causes them to isolate themselves from the community and to look to one another for support (p. 44).

Vernon Fox (1983) in his book *Correctional Institutions* offers a similar perspective with respect to the corrections context: "The staff tend to develop loyalty to 'prison people' to insulate them from intrusion or criticism from the outside community…." (p. 162).

We should, of course, keep in mind that one of the great benefits of community policing may be that the gulf between the police fraternity and the general public is not as great as it might be under other modes of policing. What do you think? If the gulf between police and the community could be narrowed by community policing, how significant a benefit would you consider this to be?

A SOBERING THOUGHT

One of the Toronto Metropolitan Police Force's deputy chiefs resigned in the summer of 2000. His name was Bob Kerr and Deputy Chief Kerr's retirement was not without some drama. For quite some time the Deputy Chief and the officer union were at loggerheads, the primary issue being one of civilian accountability. It was the deputy chief's opinion that if police officers have nothing to hide that they should not have a problem with civilian accountability. The hostility between Kerr and the union grew to the point where the union, it was alleged, was attempting to blackmail him into retirement with some unspecified "dirt." They found no dirt.

Ms Susan Eng, former chair of the Toronto Police Services Board, said of this situation, "He stands out in this city, out and above the fray, someone with intestinal fortitude willing to go the extra mile for his convictions. Some use the word traitor, that he betrayed the code: 'Don't ask, don't tell.' *It's not the betrayal that's the problem, it's the code."* (Emphasis is mine) (*National Post*, 5 May 2000, p. A21).

In an interview, Kerr told the *National Post*: "*the code has always been, and will always be, part of policing.* My daughter is a police officer and I can see it still. You have to get along with each other, it's a teamwork approach. I understand all that aspect, it's just when you get to those cases where you think: *'Somebody should really be telling the truth here.' It's tough"* (Emphasis also mine) (*National Post*, 5 May 2000, p. A21).

These are sobering thoughts, aren't they?

SUMMARY

This chapter continued to explore the discussion on officer subculture that we started in Chapter 5, by incorporating and considering the perspectives of correctional and police practitioners.

Their perspectives confirm the idea of officer subculture as one characterized by an unwritten, informal code of conduct influencing, if not governing, the conduct of officers at work. Chief among the norms of the code, as is the case with the criminals' code, is the requirement to remain loyal to one's colleagues. The practitioners all talked about the probable and unpleasant consequences of disloyalty.

Links between officer and criminal subculture were confirmed. In addition to the loyalty norm, both groups tend to see the sex offender on the lowest rung of the offender ladder. A shared argot also appears to exist.

The practitioners confirmed that the officer subculture requires identified "enemies" for its continuing existence. These enemies were identified as the criminal element, the "bosses," and the public.

Reflections

Would you agree that members of a subculture are likely to bind together more closely if they have identified "enemies"?

What are your views on the discussion in this chapter with respect to officer subculture being anti-inmate, anti-administration and anti-public?

Is there a place for officers holding each other accountable for their behaviour?

What techniques can you think of to help you keep a sense of perspective about members of the public—that is, to help you avoid seeing virtually everyone other than your colleagues in negative terms?

chapter seven

Tough Decisions

The way I see it, the bottom line for making decisions when there is peer pressure, is to ensure that you are happy with the decision for yourself, not for others.
—Journal entry. Justice student in ethics course.

When making an ethical choice, one must consider the truth versus loyalty, the individual versus the community, short-term versus long-term, and justice versus mercy. The bottom line, though, is that you have to make your own decision because you have to live with it.
—Journal entry. Justice student in ethics course.

Once we get into the field, we will depend on the money, that is the nature of our society, and we will justify almost anything to maintain that paycheque. How will you explain to your wife and kids (not to mention the bank) that you quit your job because of an ethical or moral decision?
—Journal entry. Justice student in ethics course.

To have the bravery to challenge one's convictions is necessary for the evolution of a person's psyche, soul and personality. But the process is painful and we all tend to shy away from it.
—Journal entry. Justice student in ethics course.

CHAPTER OBJECTIVES

After reading this chapter, you should be able to

- Explain how subcultural pressures can sometimes affect the choices officers make
- Explain William Tafoya's concept of the **vortex**
- Explain how the roots of officer wrongdoing may reside in allegiance to the solidarity norm
- Identify two possible ways in which an officer's moral choices can adversely affect his sense of personal and professional well-being

THE PERILS OF UNTHINKING LOYALTY

Permit me to stick my neck out a country mile (as if I haven't stuck my neck out already!). I suggest to you that the solidarity norm of police and correctional officer subculture constitutes fertile ground for officer wrongdoing. I am *not* suggesting that because police and correctional officer subculture is characterized, among other norms, by the solidarity norm, all officers are therefore wrongdoers. As we have suggested in earlier chapters, individual officers have found ways to maintain a sense of personal autonomy in spite of the subcultural norms. There is little doubt that most officers are honest and professional in the way they carry out their responsibilities.

What I am suggesting, however, is that it is not unusual, on those rare occasions when an individual officer or group of officers are adjudged to have acted unprofessionally, to discover that the solidarity constraint may have played some part in the initial wrongdoing, and almost certainly in any subsequent wrongdoing relating to the original incident.

Let me give you some police and corrections examples. We will look first at the policing context.

This incident we are going to discuss took place on a small police force of approximately 100 sworn officers in a small Canadian city of about 65 000 people. Three off-duty officers had attended a party. All three had been drinking. They were driving home in the early hours of the morning and the intoxicated driver, a little less drunk than his two partners, crashed the car. The accident scene (there were minimal injuries) was attended by a police officer who, of course, was a colleague of the three intoxicated officers. The four devised a "story" that the car had been driven by the wife of one of the officers, and that she had wandered off dazed and confused after the accident. The investigating officer, knowing the story to be false, nonetheless wrote in his notes that the three officers had been passengers in the car—that none of them had been driving it. The truth subsequently came out, and the officers were reprimanded. Here is a case of an investigating police officer knowing the lawful and proper action that should be taken under the circumstances but deciding, in the interests of protecting his colleagues, to do the wrong thing.

It is perhaps difficult to stand in judgment over him. What is a colleague supposed to do? To act ethically and professionally in this case might well have caused him subsequently to be labelled as disloyal, as a "rat," as someone not to be trusted by colleague officers. He found himself caught between the proverbial rock and hard place.

Another example took place in a large metropolitan police force in Canada. Let me give you a little background. In 1996, a young on-duty police officer was tragically shot and killed by a drug dealer. It was a particularly sad day for the police department, not only because of the tragic circumstances, but also because the young constable's dad was a retired detective with the same force. The sadness experienced by all members of the force, as you can imagine, was profound.

A year or so after the burial of the young officer, two police officers from the same force were out on patrol in an unmarked car when they noticed one of the occupants of a slowly moving vehicle expose his buttocks to three young women in another car. The cars were travelling along an inner-city freeway. The car was stopped, and the occupant doing the "mooning" was charged with performing an indecent act. The man exposing his buttocks turned out to be the brother of the deceased officer. The accused appeared in court and was given a conditional discharge, which meant he would not have a criminal record. He was fined $100 and ordered to write a letter of apology to the three young women.

Since the laying of the charge by the two officers, they claim to have been harassed and discriminated against. One senior officer, they claim, said to them, "He's a policeman's son, for Christ's sake," meaning, presumably, that they should have responded to the situation differently. They have since been taken off plainclothes work and put back in uniform, something they see as punishment for what they consider to be a case of their having applied the law fairly and impartially. As a result, they submitted a 40-page report alleging corruption in the force (*Toronto Star,* June 23, 24, 1999, pp. A10, A22).

This appears to be a case in which two police officers paid a price for daring to break their adherence to the police fraternity and its code. Clearly, many of their colleagues and several of their senior managers felt the right thing for them to do was to issue a warning only—presumably because the offender was the brother of a slain officer and the son of a retired and well-respected veteran of the force.

What do you think the two officers should have done? What would the three young women have felt, had the offender not been charged? What do you think they would have thought of the court's disposition? Is this one of those situations in which discretion should have been applied?

Mark Baker (1985) in *Cops* recounts an interview with a police officer who talks about how he and five other officers were called out to a domestic situation. "One kid moved over to one of the cops and said something to him. The cop pulled his gun, shot the kid in the head and killed him. Outright. Now you've got five cops standing there and going, 'Holy shit, why did you do that for?'....They all got together and they figured, 'We got to do something. We'll say the kid had a gun or a knife or....' They come up with all these stories.... Nevertheless, these cops felt that kind of bonding relationship with him to protect him. So they all lied. They didn't have to say that kid had a gun. But they wanted to protect their brother, because he was one of them" (pp. 199–200).

Here is a case of police officers, holding positions of public trust, being prepared to lie for each other, even when by doing so they knew that justice would not be well served.

One young woman recounted to me a disturbing story of a young police officer's experience. That police officer was her brother. "My brother is a relatively new police officer and he too has already felt the pressure of the officer subculture," she wrote. "He always said that if ever he became a cop that he would remain his own person, regardless of the pressures of the force. But even he, who is one of the strongest human beings I know, has fallen down when it came to making a decision with an experienced officer. They were out roaming the streets," she continued, "when they got a call on the radio about a robbery that was going down in the 'projects' just minutes away from where they were. The old-timer said to my brother, 'I'm not racing to that fucking Niggerhood, let someone else pick up the call.' My brother, who is clear of all racism, was shocked. He could not believe the officer was going to let the robbery happen because of the neighbourhood it was taking place in. My brother had a choice to let his supervisor know what this officer did (or in reality, did not do) or to stay quiet about the whole thing. My brother was one of those students who tried and tried and tried to get on a police force, and after many attempts he finally made it. He knew that if he ratted out his colleague to a higher authority he would be putting his job on the line—the job he had worked so hard for." Then she finished, "As much as my brother hated himself for doing it, he decided to stay quiet to save his credibility."

What a tough spot for this woman's brother to find himself in! The event proved to be a severe test for his character, but because of fears engendered by the subculture, he felt powerless to do what he thought he *should* do.

Here is one more example. I include it here because it speaks to how trainee police officers can be introduced to subcultural norms. A police education student, after attending a police ethics seminar I had conducted, asked if he could speak with me privately. He said he was troubled. He had just completed a three-week practicum and his coach officer was a 20-year veteran of the police force with which the student was placed. On one shift, he and the coach officer were told to go and ask some questions of a particular community resident. When they got to the house they discovered that no one was home. The house was in the country and around it were several outbuildings. The officer decided to walk around the property and look in the outbuildings even though he neither had a search warrant, nor could justify obtaining one. They looked in one of the buildings and saw a van that they discovered was unlocked. The police officer opened the door to take a look at the serial number and found that a piece of paper had been placed over it. The van had obviously been stolen. The police officer said to the student, "Did you see a piece of paper over that serial number?" The question clearly was asked in such a way that the student was supposed to reply something like, "What piece of paper?" In essence the police officer was trespassing and acting illegally, and now he was going to lay a charge of theft against the resident. But it was important to him that he could rely on the student. He said, "When this gets to court it is very important that you back me up and you say that the serial number was visible."

What if this trainee had said that he would have difficulty doing that? What if he had questioned the actions of the veteran officer? One doesn't have to be a rocket scientist, I think, to recognize that the trainee is in a difficult situation here. If he assures the coach officer that he can be counted on, he has taken the first step toward—if not down—a slippery slope. If he indicates that he cannot make that promise to back up the officer in court, then he puts his future in policing in jeopardy because he will be seen as someone who is not solid and who, therefore, cannot be trusted.

As an aside, there is a significant message here for senior officers when they select coach officers, and for coach officers once they have been selected. Rookie officers emulate the behaviours of coach officers; they quickly pick up both the good habits and the bad ones. Coach officers have a tremendous responsibility to perform at their best when being shadowed by a new officer. In a similar vein, police managers have a great responsibility to select coach officers wisely because of their potential for influencing their protegés.

Now let me give you some examples from corrections. The following account was given to me in one of my research interviews. I include it in its entirety because it delivers a poignant message.

This inmate was a real bad actor on this particular living unit where I was working at the [identity withheld] Detention Centre and he ended up there.

I watched him, and many of the staff had talked to me about him. I was the supervisor on the floor. And many of the staff over a period of days had talked to me about him and let me know that this guy was a trouble-maker. They couldn't catch him, but they knew he was causing trouble and creating a lot of tension. He was picking on some of the Black inmates; he was a White inmate himself, he was a big heavy guy, he was very strong, he was very muscular, and it looked like he controlled the unit.

One day I was walking past the corridor and I saw him cleaning outside the cells but in the day area where inmates would normally come out and watch television, play cards, and listen to the radio. All the other inmates were in their cells, but this particular officer had this guy out sweeping and mopping. This guy went to the cell of one of the Black inmates, the inmate who was in his cell was no problem to us at all, he was very cooperative, doing his time, awaiting his trial, but he was also very big, about the same size as the other character I'm talking about. The cleaner went up to this guy's cell, the inmate inside the cell came to the grill and the cleaner—I watched the whole thing myself—spat at the black inmate on the inside, but there was plexiglass in between so it trickled down the glass. I immediately called for backup.

I had already made the decision this was inappropriate behaviour, there was no cause for this from my observation, and I was moving him to segregation. You understand that I had lots of information prior to this. The staff had serious concerns about him and the tension caused by him. It was one example for me that I could use to put him in segregation, to take him out of the unit and maybe solve some of the tension. So I called for backup and about seven staff came to the living unit. I told them to wait around the corner and I went with one officer who unlocked the doors for me and I called this guy out, and he said, "Where am I going?" and I said, "I want to talk to you out here." I eventually coaxed him out into the corridor and he said, "Where am I going?" and I said, "You are going to segregation because I just saw you spit at that guy in his cell." He threw these big arms up and said, "There's no fucking way, I'm not going anywhere," and I thought he was going to hit me because his arms had come up and then I lunged with my hands to grab his arms and lean against him on the glass. Just then all these seven other people came flying around the corner and the dance was on for them. What they saw, it appeared to me, was an opportunity to take out all their historical anger from the last two or three days, and they were kicking him, they were punching him, and they were pulling his hair.

It was, in my estimation, a very clear case of assault versus restraint. All they had to do was restrain this man but the vented anger unloaded and it was all I could do as a supervisor to stop it. I was yelling and screaming, I was grabbing legs and pushing people back, and it was a very uncomfortable situation for me. This guy was beaten up and I got him to segregation, left him in his handcuffs and leg-irons for a little while before I would go in and talk to him, because he was pretty angry. He didn't report it. I didn't report it as a physical assault.

This officer then talked about how he felt about the situation and the role he had played in it:

Was it unethical? Absolutely. Did I do anything about it? Absolutely not. He was placed on a misconduct and in his adjudication he didn't say anything about it, I didn't say anything about it, so the facts of the intervention never surfaced.

This is an example of somebody—me in this case—of somebody who bought into the officer subculture and didn't want to get seven staff dragged through [sic] a carpet of suspensions and investigations. Now had the inmate said that he wanted to report it [the abuse] I think I would be here telling you a story about someone who had lied, who heard the inmate's allegations but went and defended himself and seven other people. It [the subculture] is a very powerful pressure.

Details of another incident were relayed to me by a justice educator who had worked for many years in correctional institutions. He said, "There was this situation where this inmate came in, was being admitted to the institution and this inmate was being very abusive to the officer, was very antagonistic, okay, and the officer lost control. The officer assaulted the inmate. The officer came to me and said what he had done. And I said to him, this is what you are going to write up." I asked the educator if the reports they wrote were fabricated. "It was a lie," he replied. He then said he had punched the officer to make it look like he had been attacked by the inmate. In this way the cuts and bruises on the inmate could be explained: there had been an altercation. "When I took that officer into the elevator, I said, 'You know what, we are going to make it look like this inmate attacked you. Close your eyes and I am going to punch you.' And that is exactly what I did."

In a young offender correctional centre, one new officer recounted the following experience: "A colleague was telling another staff member and me that he had banged a resident's head off the wall because the resident was being physically aggressive towards him. He had banged his head off the wall while trying to put him into a restraint. He had done it on purpose. He had documented it as an accident. The other officer was laughing, and I was totally shocked and had a stunned look on my face. I also couldn't believe the other officer had laughed about it. I couldn't say anything except ask him, 'What did the resident do when you knocked his head off the wall?' And he replied, 'He just cried.' That was all I said to him and my conscience is eating me up. I don't trust him around the residents while in rough situations because I know he would hurt them out of anger and would not care." There was a moment of quiet, and then she said, "I would rather not say anything because I don't want to cause any trouble because I just got hired there. I know this is sad but I am telling the truth."

This is a tough situation for a new officer to find herself in, but it is not an entirely uncharacteristic experience in corrections. The officer in question made a decision not to respond to the event she had just experienced, although she made it clear to me that she did not feel good about her decision. The desire to fit in and be accepted was simply too strong for her at that point in her career.

In the policing examples, I included one incident involving a trainee police officer. The following similar example involves a trainee correctional officer.

An inmate was admitted to a detention centre to await trial on charges of having had sexual intercourse with a two-year-old child. One of the officers decided this was such a despicable offence that he was going to exact some measure of revenge for it. He went into the inmate's cell, picked up the Gideon Bible and hit the inmate in the face with it. The inmate's eye was cut, and subsequently swelled shut. Several officers, including the trainee officer, were in the vicinity and witnessed the assault. Later when a lieutenant was investigating the incident, each officer, including the trainee, were asked what they knew of the incident. They all knew what had happened but denied having any knowledge. The trainee later recounted to me that the decision for him was a no-brainer—there was no way he was going to tell the truth because it would have been too costly for him. In all likelihood, his future in corrections would have come to a grinding halt had he broken ranks and told what he saw.

One of my students, who completed his four-month field practicum in a maximum security correctional institution, described in an ethics class an experience he had had. He was standing in a corridor mainly housing Black inmates. He overheard one of the officers telling one of the inmates that he was a "fucking nigger" and that "he was no better than a monkey in the zoo." The inmate group, as one might expect, erupted. The commotion drew the attention of the sergeant, who went into the range to quiet down the inmates. According to the student, the sergeant first whispered something into the ear of the officer, and then proceeded to ask the inmates to be quiet. He appeared to chastise the officer for what he had said. The inmate group quieted down and the sergeant and the officer left the range together. Outside, and within earshot of the student, the sergeant said to the officer, "You're right, they are a bunch of fucking monkeys—I had to pretend to be telling you off in there or they would never have been quiet. Let's forget the whole incident."

The agony of the student who witnessed this incident still haunts me several years later. He was deeply disturbed by this incident but chose to do nothing about it. In a journal written during the ethics course that I was teaching, he wrote, "I know that I should have spoken up to the lieutenant on duty and told him that I heard the officer make the two racial statements. However, I hesitated because I was scared that it would jeopardize my hard-earned respect with the officers. I am greatly disappointed with myself for not doing what my conscience wanted me to do. I feel that I have contributed to the racial problem by not doing what I knew was the right thing to do."

While I was preparing the second edition of this book, a college ethics teacher sent me an e-mail, "Why we do this job." The teacher had been using *Reputable Conduct* with a class of justice students and had received an e-mail from a member of the class, which he then forwarded to me. I sought permission from the student to include the message in this edition because it conveys the loneliness and agony of a decision-maker facing a dilemma:

> Talking about de-individuation in class and the potential for police/corrections officers to "lose themselves" and compromise their moral autonomy over making moral/ethical decisions seemed very distant to me. I mean, we're not police or correctional officers yet.... It doesn't have a direct impact on our lives at the present time. And I know you have told us time and again that the purpose of this course is not to drive certain morals and ethics into our lives but to make us think. It didn't hit home until this week just how much thinking will be involved, and how difficult the thinking process would be in situations where such a problem would arise. It seems now, more than ever, that the process we are undergoing and the concepts that we are learning are applicable to us here and now, and the situations that we face are just a starting point on a journey that will be taken for the rest of our lives.
>
> I have volunteered with [name of police force] for the last nine months. Seven of those months were on a full-time basis (8–4, Monday–Friday, and some weekends). For them I do a variety of jobs ranging from court work to community-related policing functions. At the beginning of my work with them I did a lot of work with the detachment's stenographers, typing court documents and preparing Crown briefs. I worked alongside two very talented individuals who guided me and provided me with the opportunity to learn many principles of working in a law enforcement environment. One of those individuals, however, was ostracized by the officers and the administration because she didn't really fit the role necessary to work in such an environment. Even that was obvious to me, having worked there for only a few weeks at the time. The officers and the administration would talk about her behind her back, openly criticized her work

(that deserved no criticism at all) and generally would go out of their way to make her feel bad some days.

This was apparent to me, but being the new guy I didn't feel it was my place to say anything. About two months into my stay, this individual went off work on stress leave. She would not return to work for five months. Upon returning to work she would last only another three days before having to go off on stress leave again. This individual proceeded to put in 43 complaints against the conduct of the administration, the conduct of officers in general, and even cited misconduct on the part of individual officers in her report.

Now this is where I come in. The complaints have led to a formal investigation on the part of [a senior officer from the head office]. I have received two phone calls and countless e-mails from officers and staff at the detachment warning me of the investigation. I was told numerous times that NO-ONE supported this person's claims. And that "I'd better not say anything if contacted," as if to cover up the obvious mistreatment of this employee. The investigator contacted my residence on Tuesday afternoon, however I was unavailable because of exams.

Wow…what a thing to think about. Now I know the right thing to do would be to tell the investigator exactly what happened, honestly and truthfully. However, I have become great friends with all of the officers to the point that I feel like an honorary member of their "subculture," if you will. I have been told on numerous occasions by many of the officers that they view me as a colleague and friend, not just a volunteer/student. I really don't want to do anything that would jeopardize that relationship which has taken so long to develop. (Who knows, someday these friends may be my superior officers.)

I'm not sure what to say to the investigator, however, I do know that whatever I say will be a very difficult decision to make. I am now starting to understand the thought process you were talking about. And I find myself applying the ideas that you taught us in class. I understand the background and the theories behind the material, but I'm just starting to grasp the thought process, and the understanding of myself that will be required in order to make these decisions and work in such an environment for the rest of my life. Little did I think that those situations would creep up so soon in my life.

Another of my students, after completing his field practicum at a federal institution, wrote the following. I think it sums up the issue of pressure to conform and adds an air of reality to the predicament officers may find themselves in. He wrote,

> I feel we are at work to do our job properly, so we can pay our bills and provide for our family. If someone else is doing something wrong, that isn't any of your business, I think a person should stay out of it.… I just feel that a person has just too much on the line to rat someone out and risk losing your job. Whether it is the inmates or the staff, no one likes a rat. If people find out you are a rat, nobody will trust you and chances are nobody will want to work with you either. If you go to work every day knowing that nobody likes you, then your day at work will not be fun. Before my field placement at [name of institution] I thought I would do the right thing every time even if it cost me my job and my co-workers' trust. Now my view has totally changed. I have witnessed the staff subculture and it is very strong. There was an officer at [name of institution] who was a rat and nobody liked him. When he came to the post there would be no conversation, and he would usually go to the computer room to be away from everyone because he probably felt very awkward. I don't think I could deal with that for 40 hours a week for a lifetime. Even though I may sound like a heartless bastard I am being realistic because I want a successful future for my family and me.

The pressure to conform, as one former officer put it, is a very powerful pressure.

These are examples of tough situations for officers, whether they are trainees or veterans. How can we expect officers to be their own persons under such circumstances? Might we be expecting too much of them when we say that we would like them to do the right thing? We shall return to these difficult questions later in the chapter.

At the time of writing, there is a "joke" making the rounds of the corrections system in my province—one that one of my students told me is related to all new staff and field practicum students. The joke carries with it a message that seems to fit the examples I have given.

Question: How many correctional officers does it take to push a young offender down the stairs?

Answer: None. He fell.

Now it would be inappropriate to suggest that the kind of brutality depicted here in any sense characterizes the behaviour of correctional officers or police officers. Perhaps this is an illustration of the kind of black humour that often marks the conversation of correctional officers. Nonetheless, there is an implied message here—troubling as it is—that all officers, including trainees, on occasion witness an incident but report something different, in the interests of protecting each other.

These are just a few examples of how our judgment can become obscured, and how the moral choices we make can be seen—at least retrospectively, if not at the time—to be faulty.

With respect to incidents where excessive use of force is the issue, one officer I interviewed asked rhetorically, "Are you holding that guy down a little bit longer to gain approval from your fellow workers? Are you using excessive force for that approval, for that acknowledgment, 'You're all right, you're one of us'?"

These are searching questions but it is important to ask them, for they centre on aspects of our value system concerning how we treat other human beings.

Reflections

If you had been the first police officer on the scene in the example of the three police officers who had crashed their car, how do you think you would have dealt with your colleagues' request to help them out of the "jam"?

"Rookie officers emulate the behaviour of coach officers; they pick up very quickly their good habits and the bad ones." What are your views on this statement?

All of the examples related above involve officer wrongdoing. Officers found themselves responding in ways that privately, perhaps, they would regret. Why is it, do you think, that occasionally we find ourselves behaving differently when we are with others than we might if we were alone?

Does the fact that you are dealing with an accused person who is alleged to have committed a particularly despicable crime mean that the normal rules of professional conduct can be temporarily suspended?

Would you have charged the brother of the murdered police officer for "mooning"? If yes, why? If no, why not?

THE MORAL STRUGGLE

In some of the examples discussed above, the officers involved actually admitted to knowing that their conduct probably could not be justified. In the other examples, where we do not know what the officers were thinking, would it also be a reasonable assumption that they knew that what they were doing could not be justified?

To think of ethics education in the justice system as an activity designed to help officers know what is right or wrong, is, in my view, a redundant exercise. There may be some isolated cases in which, either because of moral bankruptcy or the gradual blunting of moral sensibility, an individual has lost sight of what is right and wrong. Generally, however, officers know what the right thing to do is in most situations. The purpose of ethics training, then, rather than being an attempt to determine what is right and what is wrong, may be to help officers with their struggle when they experience personal moral conflict, either when working alone or in their colleague group. Such a purpose can only be realized, I think, if the ethics educational experience is one characterized by empathy on the part of all concerned—teacher and student alike. If the experience is one in which participants feel they are being judged or manipulated, this would be a prescription for failure—painful to both teacher and student.

The Potential Cost: Moral Conflict Can Make You Sick

Let me offer an idea not generally discussed in any literature that I have encountered. I would suggest that an individual officer's sense of well-being or self-esteem can be positively or adversely affected by how he responds to these moral struggles. I would further suggest that there is a link between how an officer deals with moral conflict and the stress an officer can subsequently experience. Though I can offer no empirical evidence, there is some anecdotal evidence supporting my view, arising out of the interviews I have conducted and from what several ethics seminar participants have confided to me. Additionally, one or two writers have touched obliquely on the subject. I am going to expand on my idea in the hope that it will become a topic of personal thought and class debate for you and your student colleagues.

Ethics education, therefore, can be seen in terms of its usefulness as a stress-reliever—as an aid to an officer's sense of well-being. It may also be seen as an aid to help an officer reach retirement relatively unscathed by major career incidents.

It is important that we keep this conversation in perspective. I do not suggest here that officers encounter these crucial moral dilemmas on a regular basis. These are the more significant moral struggles that may be experienced only occasionally during an entire career. But as few as they may be, they are defining moments. How they are responded to can, to a large extent, establish the tenor of the individual officer's professional and personal life. Persistent irritability, relationship breakdown, alcoholism, tarnishing of reputation, loss of colleague respect, or loss of job are just a few examples of what can happen to an officer if found guilty of responding inappropriately to a moral struggle. And any or all of these effects, if experienced, can have a detrimental effect on an officer's general state of health.

We referred earlier to researchers Elizabeth Grossi and Bruce Berg (1991), who wrote a journal article called "Stress and Job Dissatisfaction among Correctional Officers: An

Unexpected Finding." Interestingly, in the context of discussing correctional officer stress, they refer to six stressors, one of which is "that one cannot always act the way either one would choose to, or the way the public might expect them [sic] to." This suggests to me the idea of moral conflict and moral struggle. It is significant that these two writers connect this struggle with the potential for stress. They add, "Again, it may be that in order to attain a satisfactory level of peer support, and reduce conflicting pressures endemic to correctional institutions, correctional officers must compromise their personal values and interests" (p. 80). Such compromise, if experienced, would constitute a fair amount of personal stress for most of us—would it not?

One of my university philosophy teachers, Clive Beck (1991b), used to talk about how our value system and our ability to follow it can have an effect on our sense of well-being, or what he called the "good life." If we lose touch with our basic values, he suggests, our enjoyment of life can be impaired. I remember Beck differentiating between *moral values*, which he listed as carefulness, responsibility, courage, self-control, reliability, truthfulness, honesty, fairness, unselfishness; and *social and political values*, such as peace, justice, due process, tolerance, participation, cooperation, sharing, loyalty, solidarity, citizenship (p. 3). You will notice that "loyalty" and "solidarity" are listed as two of the social and political values. And worthy values they are—in the right context. But what if they conflict with other values? What if the result of embracing loyalty and solidarity as values above all others is a diminishing of individual courage, or truthfulness, or fairness, or honesty, or justice, or tolerance, or due process? If the latter values matter—and clearly they do for most correctional or police officers—there is going to be conflict and struggle, where one's decisions may well have a direct effect on one's sense of well-being.

My former teacher has a rather unusual way of looking at morality, and makes the point that it is generally unhelpful to blame people or to make them feel guilty for their choices. Rather, he sees a person's immorality as a problem for that individual, something that reduces well-being. He sees such people as unfortunate because they lack the motivation (and courage?) to do what is best for themselves (p. 11). So it would not be too much of a stretch to link the idea of ethics education as it relates to helping us deal with moral conflict, with the successful management of stress in our professional lives.

Here is a graphic example of how a person, a correctional officer in this case, can have his sense of well-being affected by an act of immorality. The example is from Kelsey Kauffman's book, *Prison Officers and Their World*. The officer is talking to the author after a brutal beating of an inmate by other officers. "Maybe I feel what I am saying on his hurt, is my hurt. Because I felt so lacking, so missing something that I didn't do something. I was totally ashamed, totally embarrassed that I had to say that I had seen, and I am really, really sorry. And I'm not sorry for him, I'm sorry for me that I didn't have the guts or the balls or whatever you want to say, to do something at that time. And it, it eats away afterwards" (pp. 233–234).

Moral Weakness or Moral Courage?

One of my ethics course students wrote a comment in her journal that impresses me for its honesty, but at the same time troubles me because it shows poignantly how she has great difficulty dealing with her moral struggles. "I know that I don't always do what is 'right' though I have a sense of what that 'rightness' is," she writes. "It is a constant source of disappointment to me, and I can't seem to rise above it. It's horrible to know what the proper moral stance is and not have the strength to take it. This is another reason for my self-dislike." Clearly, she is an example of a person who is paying a fairly significant personal price for her moral struggle and for what she considers to be her lack of moral courage. She typifies, I think, what Beck means when he says that a person's immorality is a problem for *that person.*

Rhetorically, this person then asks in her journal a searching question: "Are those who don't see anything wrong with an unethical position happier than those who know the difference, but can't take a stand?" If my former teacher is correct in his thinking, the answer to this question is yes. It is only at the point where we have a good idea about what we should do, but feel we can't do it, that the serious, health-harming conflict can begin to trouble us. The personal stress begins precisely at that point when we know what we should do, but feel that we do not have the moral courage to do it. Often, I would suggest, our courage is lacking because we fear the consequences of taking a stand. And what I have heard consistently from officers is that what they perceive to be moral failure on their part comes with a price.

One correctional officer trainee referred to this issue of moral courage when he said, "Do you stand alone and become a leper, or do you let peer pressure and group think compromise your ethical stand? This is a hard question when you have to feed your family." There is, of course, no easy answer to this question. This person's concern is a legitimate one.

These uncourageous responses, often precipitated by the very real fear of reprisal, can cause many of us (myself included) to respond to the moral struggle by taking the safest route, the route of least resistance. As a result, the decisions we often make are ones that we would prefer not to make if we were stronger. We often take the route that we think will cause us the least amount of grief with our peer group. And often we neglect to think of the private angst that may result from the moral choices we make, and how that may impact on our sense of well-being—on how we feel about ourselves as persons.

Gail Sheehy (1977), in her book *Passages: Predictable Crises of Adult Life*, sees true adult development in the achievement of what she calls "autonomy," and "self-sufficiency" (pp. 52–56).

But this is a difficult enough goal to achieve in our "regular" lives away from the job. It is a much more difficult goal to achieve for police and correctional officers who can be influenced by the subcultural constraints of loyalty to colleagues.

In Chapter 1, reference was made to the Somalia inquiry in Canada. That inquiry into the killing of a Somali youth by some Canadian soldiers on a peacekeeping mission has become characterized, I think many would agree, by what appears to be stonewalling and a conspiracy of silence on the part of many of the witnesses. In addition, many attempts have been made by the military to discredit the physician who was the first person to raise a concern about the way the youth died. The physician appears to have shown great moral courage, while many other witnesses have not. It is possible the others have not because they fear that they will have to endure ostracization and discrediting similar to that experienced by the physician. But the physician was able to summon the courage to expose what he considered to be a wrongdoing.

What enables some individuals to risk everything to take a moral stand? One writer, Thomas Lickona (1976), a social psychologist, asks a similar question: "What enables a person to maintain moral perspective in the face of strong situational pressures?" (p. 124).

I guess if we could answer that question, we would all be much stronger in the face of moral conflict. One prolific writer in the area of morality, Lawrence Kohlberg (1984), thinks part of the answer may lie with what particular "moral stage" an individual has achieved. Kohlberg talks about six stages ranging from stage 1—representing an unthinking, undisciplined moral sense—to stages 5 and 6—representing what Kohlberg calls "principled moral thought" (pp. 454–459).

Lickona (1976), in an article entitled "What Does Moral Psychology Have to Say?," refers to a US soldier in the Vietnam War. You may remember the Mai-Lai massacre, during which an entire village was burned and many of its inhabitants, including women and children, were torched by US soldiers. The soldier's name was Michael Bernhardt, and he was the only ground soldier who refused to participate in the massacre. He was found to be operating at a high stage according to Kohlberg's definition of moral reasoning, and here is a statement Lickona attributes to him: "The law is only the law, and many times it is wrong. It's not necessarily just, just because it's the law. My kind of citizen would be guided by his own laws. These would be more strict, in a lot of cases, than the actual laws. People must be guided by their own standards, by their self-discipline" (p. 125).

You will notice there is no sense of blind allegiance in this statement. When one considers what the pressure must have been like for Bernhardt to conform with the rest of his highly trained soldier colleagues, it is amazing that he was able to remain his own person.

In late 1998 I attended a police ethics conference in the United States. One of the guest speakers was Captain Hugh Thompson, who had also been involved in the Mai-Lai affair as a helicopter pilot. For many years after the war he had been castigated as a traitor. His offence was that when American soldiers were shooting unarmed villagers, he landed his helicopter between the soldiers and the villagers and ordered the soldiers to stop shooting. He went further. He announced over a loudspeaker that if they did not stop shooting, he would order his men to open fire on them.

Many years later, Captain Thompson is being hailed as a hero. The documentary program *Sixty Minutes* did a story on him, and reporter Mike Wallace took him back to Vietnam, where he met with some of the villagers he had helped to save. The Pentagon has also recognized him as a hero for the role he played.

Captain Thompson, in his speech at the conference, appeared to be a self-effacing individual. When asked why he did what he did, he responded: "I don't see how I could not have done what I did."

Both Michael Bernhardt and Captain Thompson showed a rare courage. They somehow knew what it was they should do, and did it—without fearing the consequences.

Of course, you may ask what kind of army one would have if all soldiers went off half-cocked and did their own thing! What would happen to the *esprit de corps* so essential to a group of soldiers fighting together?

Keep in mind here that we are discussing situations *in extremis*, as it were; situations that cause a significant crisis of conscience in an individual. Generally, you can still have your togetherness and solidarity and camaraderie—but only up to the point beyond which an individual would betray her sense of morality.

I remember reading a newspaper article in Australia entitled "Why Loyalty Is a Two-Way Street." One question posed by writer Adele Horin was this: "Is a loyal Nazi better than a disloyal one?" She made the point that loyalty "can blind you to higher loyalties, to truth-telling, justice, the common good." After pointing out that loyalty to a group can blind us to what she calls "wider commitments and obligations," she adds, "For example, the Blue Wall of Silence around police is a wall of loyalty and fear that keeps buddies true to each other at the expense, sometimes, of the good name of the force" (*Sydney Morning Herald*, 9 July 1994). It is significant that she uses the word "fear" here. The fear, presumably, would be that of being ostracized by the group.

One of my current students, as I write this chapter, has a brother on a Canadian police force. His brother has told him the worst thing that can happen to a police officer in his professional life is to be ostracized by his colleagues. It is clear that it demands considerable courage to take the kind of stand Michael Bernhardt and Captain Hugh Thompson took.

But let's return to this idea that *not* being courageous can sometimes, in a real sense, have an equally devastating effect on us and our careers. William Tafoya (1995), in an article "Ethics and the Realities of Life: Surviving the Vortex," graphically depicts how one police officer got caught up in what Tafoya calls a **vortex**: a kind of whirlpool that swallows or consumes everyone in it. The incident he refers to is one in which a group of police officers in a major metropolitan police department were terminated for their part in the beating of a Black motorist, and the subsequent cover-up. One of the terminated officers spoke with Tafoya about the event several years later, and much of the article is a verbatim account of the interview. The article is a gripping, and sobering, account of what can happen to a good, but uncourageous, person.

In his introduction, Tafoya expresses something amounting to incredulity that an otherwise intelligent officer would feel compelled to lie. This was the case despite the fact that the officer had come from a solid upbringing in which values and doing the right thing had been highly prized, and the fact that he knew what he was doing was wrong. Tafoya points out that even though this officer felt the cover-up was going to fail, he "nonetheless believed he must go along with the attempt to do so" (p. 576).

We clearly do not have the space here to include the long transcript (which in my view ought to be mandatory reading for every police officer), but I do think it important to include some of what the former officer said to Tafoya. In the transcript of the interview, Tafoya notes in several places where the former officer either began to lose, or actually lost, his composure. Keep in mind that the incident had taken place several years earlier, and that he had made a relative success of his life after policing. Despite this, it is clear that he had suffered severe emotional and psychological pain at the time of the incident and in the ensuing years. Here are some of the things the former officer said:

> This goes back a long way, before I joined the police department. I was socialized early and often in the police culture by friends who were police officers. From them I heard repeatedly that the most important lesson to be learned was sticking together. One friend [Fred] suggested that if I ever saw anything that I thought was improper—and was a rookie—I should keep quiet, go back to the car, and make like I was writing a report; stuff like that…. A woman in our training academy class had previously done an internship with the department's Detective Bureau…. We asked her about what it meant [the expression "hand-up guy"]. She said rather sarcastically that "hand-up" is the worst insult one police officer can give another. A "stand-up" cop is one who covers for a fellow officer. Everybody likes such officers, trusts them. Such officers go to bat for one another. But a "hand-up" cop is likely to tell the truth and not cover for a fellow officer. It was then that I first began to understand that in the police culture, telling the truth was not always a virtue.

You will notice that these sentiments in a real way echo much of what we have been discussing in the book to this point. Now let's take a brief look at some of the things this fellow had to say about the effects on him personally of his inability to take a moral stand.

> I grew up in a little Christian Reform community. I'm a religious person. I know what the truth is. I know you are supposed to be honest. That is probably what has bothered me more than anything else. I know what I did was wrong. I knew it was wrong when I did it. I was sick with myself; still am.... To this day I struggle with this.... Oh God [becomes very emotional]. What I did, I'm still ashamed of what I did. A comment a good friend made still haunts me. He said, "Integrity is like virginity. Once you lose it, it's gone" [begins to cry]. It's been a lot of years; that's an awful long time. I don't think as though I've got closure yet. I don't think I am going to wake up and find out it was all a bad dream. It happened and I'm still having problems with it (pp. 577–580).

In my view, this is a graphic and disturbing account of how an otherwise principled officer, in the interests of being seen by his colleagues to be a "stand-up" officer, was prepared to put at risk, if not lose, everything he valued in life, including his reputation and career. For many of us, such an experience would certainly result in a health-harming loss of self-respect. Indeed, at one point in the interview the former officer said that the primary reason for his agreeing to the interview was to try to come to terms with the whole issue. "That's why I'm letting you do this to me [the interview]," he said. "This is a catharsis for me. I've gotta stop somebody else, some other police officer, from doing this. I don't want somebody else to go through what I did" (p. 583).

Of course, one big question that has yet to be addressed is this: What if we make a decision that to be ostracized by colleagues would be more stressful than anything else that can happen to us? This is a question that needs to be asked, and one we will address in Chapter 8. The former police officer in Tafoya's interview was asked what police departments could do to help officers like him who essentially open themselves up to becoming corrupted by responding in inappropriate ways because of subcultural constraints. He described his dilemma thus: "I guess I was damned if I did, and I was damned if I didn't" (p. 580).

One final thought on the discussion in this chapter specifically, and in the book generally up to this point. There have been times during the process of writing that I have felt compelled to complete the book because of certain "events" that occurred in my life. One such event happened two days ago. I had just returned from a two-week trip overseas, and there were several messages waiting for me on my voicemail at the college. One was from a former student who is now working in corrections.

> I am phoning to apologize to you, John, [began the call,] because when you taught us the ethics course I did not accept what you said to us that there may be moments in our career when we would find it very difficult to make the right decision. I always felt that I would never experience such difficulty because I was convinced that I would always be able to do what I thought was the right thing, that there would be no question about it. Well, I'm phoning to tell you I was wrong, and I want to apologize to you for not believing you. I am now confronted by a situation at work where, if I do what I think I should do, I will lose my job. So I just wanted to apologize to you and to say you were right in what you said in class.

Now it doesn't matter to me whether what I said in class was "right" or not. I was simply attempting to get the students to think about issues of professionalism and moral responsibility. What I did find upsetting was the tone of voice of the student. She did not leave a phone number, and I do not know where she is working, so I have been unable to return her call. She is clearly struggling with the problem in a way that is adversely affecting her life.

The personal price we pay in attempting to deal with our moral struggles can be significant. This is where, in my view, ethics education approached in an empathetic way may be of some help.

Reflections

Would you agree, for the most part, that officers know what they should do in most situations, and that ethics education geared to teaching what is right and wrong might be a redundant activity?

What is your view of the idea that our sense of well-being as individuals is connected to how we resolve our moral struggles? What is your experience?

Have you ever considered how you would deal with a situation in which loyalty to colleagues might conflict with the values of justice, fairness, or honesty?

Why is it, do you think, that individuals such as Michael Bernhardt are able to take a moral stand even when the consequences are likely to be painful?

Is a loyal Nazi better than a disloyal one?

A friend of the former police officer interviewed by William Tafoya said, "Integrity is like virginity. Once you lose it, it's gone." How do you react to this statement?

UNION SOLIDARITY

We have been discussing how difficult it can be to make courageous personal moral decisions. We don't wish to be seen as different and we certainly don't want to experience what it is to be an outcast.

Now it would not be unreasonable to suggest, and there certainly is anecdotal evidence to support this idea, that police forces or correctional jurisdictions that have very strong, tight unions are likely to experience even more subcultural constraints on its members than those agencies where the union is not as unified or powerful.

It is important to acknowledge that unions or officer associations can serve a very useful and important function. Line employees can be managed poorly and/or unfairly and it is good that they have their union to support them at such times. Ensuring its members are fairly compensated and enjoy working under reasonable health and safety conditions is a very important function of a union.

But what if your union is acting unwisely, improperly, or unfairly in your view? What if the union insists on representing aggressively its members who file petty, stupid grievances? What if, in fact, your union actively *solicits* petty grievances from its members just to get under the craw of the administration? What if your union makes what you think are unfair demands of a reasonable and principled administration? And what if your union calls for a strike that you think is an ill-considered and unreasonable one? What then?

In 2002, there was a widespread public service strike in Ontario, Canada. It is interesting that the most militant sector in the call for a strike were those members who work in the justice field. The strike was a long one and much bitterness ensued.

One of the justice jurisdictions involved in the strike was the scene of particular tension and bitterness, and officers have told me that the tensions that developed between colleagues will, in their view, never dissipate.

This state of affairs arose because some members were more moderate in their approach to the strike and as a result were seen to be less solid than they might have been. At one point during the strike the following words were painted large on a road leading from that agency:

SCABS CAN'T HIDE

Severe threats were made against individuals and some feared for their safety.

The question is this: Should you have to give in completely to the leadership of your union when you feel that leadership is acting irresponsibly or unfairly? Isn't it interesting that we label conscientious dissenters "scabs"? Now I am not suggesting here that active union membership conducted reasonably and fairly is not a good thing. I think it is a vital and necessary part of any public service agency. But what can or should we do when the leaders seem to be intent on causing nothing but trouble and grief? Only you can answer that question. It's a tough one, and it's an even tougher question when those leaders appear to enjoy the support of the majority of your colleagues.

SUMMARY

In this chapter we have suggested that the solidarity norm in officer subculture constitutes fertile ground for officer wrongdoing. We discussed some actual incidents, from both corrections and policing, to illustrate how this might be the case.

In the discussion, it was suggested that officers, for the most part, know what is right and wrong, and that ethics education that is geared toward helping individuals know what is right or wrong is probably redundant. It was suggested that the purpose of ethics education should be to help officers cope with their moral struggles. As a result, officers on occasion may be able to avoid falling into a trap—or, to use Tafoya's term, "vortex"—similar to that in which the former police officer interviewed by Tafoya found himself.

One of the benefits of ethics education is that, directly or indirectly, it can contribute to an officer's sense of well-being. That is to say, if an officer, as a result of an ethics education experience, manages to deal with her moral struggles in a personally satisfying way, her sense of self-esteem will be maintained—or even enhanced. Of course, the converse would be equally true, as was so dramatically illustrated in the case of Tafoya's interview subject.

chapter eight

Tools for Moral Decision-Making

I have always had the notion that we never truly think while in college. All we do is memorize.

—Journal entry. Justice student in ethics course.

The answer to a particular issue or problem isn't the important thing; the important thing is how we arrive there.

—Journal entry. Justice student in ethics course.

The first roadblock I encountered on my ethical journey was leaving my emotions and feelings at the door...I slowly realized how cloudy an issue becomes when you include your emotions and feelings.

—Journal entry. Justice student in ethics course.

To me, ethics is what we choose as right or wrong, how it compares to what society decides is right or wrong, how we choose to react to the differences, and finally, what we choose to do when we are confronted with the conflict of what we should do.

—Journal entry. Justice student in ethics course.

I feel that it is very important to look at things critically. Far too often people take things at face value, for example, as in information presented in newspapers or on TV. I think we need to consider where the information is coming from, who is saying it, and what is really being said. By thinking critically about things, I feel we can become more educated—and safer.

—Journal entry. Justice student in ethics course.

CHAPTER OBJECTIVES

After reading this chapter, you should be able to

- Explain why dilemmas are usually complex
- Explain why it is that no matter how we respond to some dilemmas, we will feel that we are in a "no win" situation
- Explain what is meant by "The choice really may be one between 'self-respect' and 'colleague respect'"
- Identify the two ideas that some officers have expressed that may help to explain how some officers deal with dilemmas they experience on the job
- Identify the seven "tools" that may help our decision-making when we are faced with a moral dilemma
- Explain why there may be a gender difference with respect to individual officers "setting limits"
- Explain what is meant by **setting limits**
- Explain the principles of **benefit maximization** and **equal respect**
- Explain what we mean by the **bell**, the **book**, and the **candle**
- Explain what we mean by the **imaginary video camera**
- Explain the general idea of **critical thinking**
- Identify five elements of critical thinking
- Explain how thinking about moral stages may help our moral decision-making

INTRODUCTION

By way of introducing this discussion on the tools of moral decision-making, consider two related ideas raised in the preceding chapter, one in the form of a question and the other in the form of a statement.

Question: "What if we decide that to be ostracized by colleagues would be more stressful than anything else that can happen to us?"

Statement: "I guess I was damned if I did, and I was damned if I didn't." [This was expressed by the subject of William Tafoya's interview when describing the dilemma that led to his wrongdoing and subsequent dismissal as a police officer.]

The intrinsic nature of any dilemma is that it has no simple solution. If the issue confronting us is truly a dilemma, then making a decision about it is probably going to be difficult. Sometimes even choosing to do what we consider to be the right thing does not necessarily mean that we are going to be completely at peace with our decision.

In particular, this would be the case if we elected to make a decision that could in some way implicate a colleague officer. I think this is the kind of dilemma with which Tafoya's former police officer subject found himself confronted. If he had made a decision not to get caught up in the lying and fabrication, then in all likelihood he would have had to deal with the subsequent wrath and rejection of his colleagues. He would probably not be trusted again, and would have to endure the anguish of being labelled a "hand-up cop," and all that can go with that kind of labelling. On the other hand, having made a decision to stay solid, he experienced not only the public indignity of being fired as a crooked police officer, but also, for many years, has had to struggle, day in and day out, with his own personal demons arising from the incident.

NO-WIN SITUATIONS

Some situations can appear to be, in a very real sense, no-win. On the one hand, you can choose to stay solid, to be seen as a "stand-up cop," and continue to earn the acceptance, loyalty, and support of your colleagues; on the other hand, you can choose to act according to the dictates of your conscience if it is leading you in another direction, and run the distinct risk of losing the acceptance, loyalty, and support of your colleagues.

It would seem that the difference between the two choices is that you can opt either to continue to be admired and respected by your colleagues but lose some or all of your self-respect, *or* you can opt to hold on to your sense of self-respect but lose the respect of your colleagues.

Of course, there may be many occasions when the retention of both self-respect and the respect of your colleagues would be the result of your decision. I am not suggesting that all hard choices necessarily result in the "either this...or that" scenario. But there will almost certainly be times in your career as an officer when you will be confronted with the kind of dilemma we have been discussing here—with a choice that really is one between self-respect and colleague respect. And when these situations arise, they can, as we have discussed, be extremely difficult to resolve, and result in personal anguish.

It is at these moments that one's sense of self—who one really is as a person—is put squarely on the line.

TWO WAYS OF DEALING WITH MORAL CONFLICT

I have had some interesting conversations with officers about this topic. They have expressed two ideas that could provide interesting material for your personal thought or for discussion with your colleagues. Some of you may think these ideas worthy of exploration, and I would encourage you to do some further research on them. We don't have time in the context of this book to discuss the ideas in detail, nor am I going to offer any comment on them, but here they are:

First, a few officers have suggested to me that some officers are able to cope with dilemmas that have the potential to put them at odds with their colleagues, simply by regarding loyalty to and solidarity with colleagues as the *pre-eminent* value in their professional lives. In other words, deciding that to abide by this principle is the ultimate morality. This one-dimensional thinking, of course, can enable individuals to get out from under some moral struggles that are essentially multi-dimensional in nature, and which do not offer a simple solution. It may be an effective way of avoiding anguish for some officers. Whether or not this approach would count as an intellectually honest one is a moot point.

Second, some officers have suggested that when they are at work, because of the unusual nature of their job and the character of the socially deviant "clients" with whom they normally work, they can mentally and emotionally justify *suspending* the morality they operate by in their private lives. In other words, they can rationalize allowing themselves to be governed by a different, possibly questionable, morality on the job. (This sentiment has been expressed almost exclusively by male officers. Remember our discussion in Chapter 6 about this topic.) If, indeed, this rationalizing does happen, it would seem here that one's moral sense is compartmentalized—that one's "real" moral sense is practised and adhered to in one's private life, where one remains committed to truth-telling and fairness, and so on; and that the other moral(?) sense in the workplace allows one to engage in what I call "means-end" morality (an individual convinces himself that even if the means are dubious—or even unlawful—if the end is a good one how you get there doesn't matter), where the normal tenets of morality can be suspended or ignored. (You will remember that we already considered this kind of thinking in our discussion of utilitarianism in Chapter 2.) I trust you will have some fun debating and discussing these ideas.

Reflections

What is your view of the idea that to be ostracized by colleague officers for taking a moral stand is one of the worst things that can happen to an officer?

If you were confronted by what you thought was a no win situation, what factors would probably influence your decision-making? Attempt to "anchor" your response to a specific situation.

Is it reasonable to talk about "self-respect" and "colleague respect"? Or does this way of thinking simply not recognize that, for the officer, the two are invariably linked?

What are your views on the two ideas expressed by officers to explain how some deal with moral conflict?

Let's return to the topic under discussion. We have suggested how extremely difficult it can be to be confronted with a no win situation. We are now going to discuss some "tools" we can use in our moral decision-making. We should come at those decisions understanding that there is seldom a perfect solution, and that there is probably going to be some pain, whatever we choose to do. The tools will enable us to make decisions that give us the greatest degree of comfort.

I guess one test of a decision we are wrestling with might be to reflect on it at three in the morning on a night of fitful sleep, and see how we feel about it!

If by using one or more of these tools you are helped to make the best possible decision for you and for others affected by that decision, then the time and effort you commit to considering them will be more than justified—at least in terms of your peace of mind and sense of well-being. More important, perhaps, thinking in this way may help, at some point, to save your career.

Now, it is true that sometimes in your moral decision-making you must make the best of a bad job, as it were; that is, make the best decision you can under certain circumstances. You need to be aware that even though you may at times come to a certain and compelling decision about the "right" or ethical thing to do, if you *act* on such a decision, you may still experience pain and anguish because of the way others respond to your choice. Taking the moral high road can be costly.

Think about these tools. Apply them in your daily life. You may find them useful. And you may find someday that they help you avoid a situation you might otherwise regret!

SEVEN TOOLS

Setting Limits

This is a technique that some officers have told me helps protect them from potentially compromising situations. They simply indicate to colleagues, in a pleasant and courteous way, what their personal limits are. They may do this, for example, regarding using too much force, or gossiping, or sleeping on night shifts, or accepting gratuities, or whatever the case may be. These questionable practices are simply cited as examples, and are not meant to imply that any one of them or a combination of them normally characterizes the behaviour of officers.

Setting these limits and making them known to colleagues can, I am told, cause a colleague who might be contemplating responding unethically to a given situation to have second thoughts about doing so in your presence.

An example might be a situation in which an officer partner on night shift tells you he wants you to "spot each other off" as you alternately take turns to sleep. You might say to him, by way of establishing your own limits, "I don't feel safe when either one of us is sleeping, so I would prefer it if we both stayed awake." If he tells you to take a running jump, so to speak, you may then say, for example, "I will not sleep, and I would prefer it if you did not. It is not safe. But if you continue to sleep I may be left with no other option but to take it further. I really do not want any unpleasantness between us, but the choice is yours." To set our own personal limits in this way can be difficult for us, but it can save an awful lot of grief later.

Perhaps it should be added here that many of the officers who have advocated this technique have been women. My research shows clearly that female officers, more so than male officers, use this technique effectively. We can hypothesize why this may be the case, and I hope that you will give some thought to the issue. You should be aware that several of the female officers I have spoken with tell me that they set limits in a variety of ways, and that it is possible to do so without alienating their colleagues. Some advocate using a sense of humour to get their point across, but they caution that this approach would depend on the kind of relationship the officer has with her colleague(s). I remember talking with one female correctional officer who explained to me that she was off on stress leave because she could no longer deal with the attitudes and insensitivity of her male colleagues. She explained that the last straw for her was when a male colleague said to her after she returned from a washroom break, "I hope you didn't get your 'hair' wet." Now, this is clearly a reference, and an embarrassing one for this officer, to female genitalia. She decided she could not put up with this kind of "humour" anymore, and cited her male colleague for harassment.

I discussed this particular incident with another female correctional officer from a different institution, and asked her what she thought of the comment and the female officer's response to it. She confirmed that the comment was clearly out of line. I asked her how she would deal with a similar situation. Interestingly, she said she would in all probability use humour.

She said she would tell the offensive officer a joke, and the joke runs something like this. God had made Adam and Eve, and was in the process of handing out male and female genitalia. He said to them both, "I have here a penis, and whoever decides to choose it will be able to pee standing up." Adam responded, "Oh, God, please let me have that one, please, please let me have it." God looked at Eve and said, "What do you think?" and she said, "Oh, if he wants it, let him have it. He'll only sulk if you don't let him have his way." God said, "Okay, then, Adam, you can have it. You, Eve, will get the other one. This one will allow you to have multiple orgasms."

The officer recounting this story explained to me that this kind of humour has worked well for her in her 20-odd years as an officer. She is able to hold her own when male colleagues behave inappropriately, and they have learned to curb their inappropriateness accordingly. This officer said she uses the same technique to give her colleagues a clear message when she is uncomfortable with something they are doing that relates specifically to the job.

The research also shows that when male officers are able to use the setting-limits technique without running the risk of alienating their colleagues, they generally enjoy the comfort of being viewed by their colleagues as credible, experienced, and reliable in an emergency. It could be problematic for a rookie officer to use this technique, especially if she is perceived by more experienced colleagues to think herself superior to them in some way. A new officer would have to set personal limits cautiously and diplomatically; otherwise, he could be setting himself up for an interesting orientation to the job! This is not to suggest, however, that if you are a trainee officer, you should not consider the advantages in adopting this technique.

Reflections

Is it realistic, do you think, for a new or relatively new officer to "set limits" with colleagues?

If you answered yes, what are the risks, if any?

If you answered no, try to explain why you feel this way.

Some male officers have expressed the view that they perceive themselves to be much more subject to the potential for what they call "parking lot justice" than are female officers. What do you think? What is your experience?

What is your view of the idea that experienced, proven officers are more able to set personal limits?

What is your view of the idea that humour may be a useful and effective way of getting a message across?

Ethical Principles

There are some writers who think—at those times when there is time to think, and I realize that this may not always be the case in corrections or policing—that it can be helpful to consider certain ethical principles in pondering a dilemma. Three university professors—Kenneth Strike, Jonas Soltis, and Emil Haller (1988)—have written a book called *The Ethics of School Administration*. In the book they skilfully explain, and demonstrate, how incorporating some thought about ethical principles into our moral decision-making can help us arrive at more informed, and probably more correct, decisions. We do not have the time to discuss the practice in depth, so I recommend that you read the book yourself if what follows is of special interest to you. The practice is what Strike et al. call "the application of principles to cases" (p. 4).

Let me caution, however, that ethical principles themselves can sometimes give us conflicting messages about what we should do in a given situation. A second caution would be that incorporating a consideration of the principles into our decision-making could potentially rationalize a decision with which we basically feel uncomfortable.

I would encourage you to practise using the principles in your own moral decision-making. One way to get used to incorporating them into your thinking is to use them as you discuss the dilemmas in Chapter 9, and any other dilemmas arising from your own experience that you feel comfortable bringing to class for examination.

The Principle of Benefit Maximization   You may remember our discussion of this principle in Chapter 2. It was formulated mainly by John Stuart Mill and Jeremy Bentham. This principle holds that when we are faced with a conflict about what we ought to do about any given moral dilemma, we should attempt, prior to making any decision, to weigh the consequences of our decision. If we judge that one particular decision, among all the ones we could make, is likely to lead to the greatest amount of good for the greatest number of people, then this decision is the one, according to this principle, that we should make. The morality of our action, therefore, is judged against the probable consequences of that action.

This principle can be useful in helping us think through our moral dilemmas. However, of the two principles we are going to discuss, this is the one that is most problematic. We suggested earlier that sometimes these principles can give us conflicting messages. The principle of benefit maximization is the principle most likely to conflict. It can, unless we are careful, be used for the purpose of rationalizing decisions and behaviours about which we do not feel at peace or, more plainly put, which we suspect are wrong.

For example, a couple of police officers may apprehend a suspect who they have a strong instinct (but no proof) is at the root of a string of crimes that have been committed recently. The officers know from their experience that this particular suspect is a bad actor and that the community generally would be safer with him behind bars. You can see already that in a case like this it would be easy to use this principle of benefit maximization to justify, for example, planting drugs on this suspect, if the officers are unable to come up with any concrete evidence. In that way, they may convince themselves they have served the community well by making it a safer place. In other words, a little bit of injustice has led, they convince themselves, to the greater good. Dubious means have led to a good end, so the wrongdoing is justified as well-intentioned and productive.

In the corrections context, correctional officers may feel that beating a convicted child molester who they feel has "got away" with a light sentence may cause that offender to think twice about reoffending, again thereby rendering the community a safer place for children.

As reprehensible as child molestation is, we should think long and hard about whether, as peace officers, we have the right to suspend the normal rules of justice, law, and order, in the course of our duties.

The thinking and behaviour in the two hypothetical situations above would be difficult to defend, either morally or legally. Many would consider the officers' responses to be illegal and unprofessional. Others would consider them to be unethical, because the actions of the police officers and the correctional officers are the sorts of actions that would normally lead to cover-ups and dishonesty in the interests of self-protection.

Their actions may essentially represent a kind of vigilante justice; the thinking that governs these kinds of actions may not be very different from the thinking that governed lynch mobs.

Let me give you another case to think about. This case involves a police force deliberately duping the media into reporting false or misleading information. In July 2002, during a murder trial, a York Region (Canada) homicide detective revealed that he had called a news conference and handed the media a fake sketch of a suspect in the killing of a woman. The *Toronto Star* published the picture the day after the news conference. In his testimony, the detective said that the ploy was "somewhat misleading, but it was not outside the bounds of the law." The detective said his objective was to stimulate conversation and activity among suspects, which he thought would give the police a better chance of gathering critical evidence.

It's clear there are issues of credibility and trust at stake here. Apparently, the detective felt that these means, however dubious they might be, were justified under this particular set of circumstances. In other words, to mislead the media might be a necessary evil in this case to catch the offender (*Toronto Star,* 6 July 2002, p. H6).

For the last decade or so this utilitarian approach (another name given to this kind of thinking) to making moral decisions has been embraced to a great extent in the Western world, particularly in North America. But because of the tendency of individuals to use this sort of thinking for the purpose of rationalizing all kinds of questionable behaviours, it is now, in the late 1990s, being questioned as a useful aid.

My view is that although it is a useful concept to keep in mind, the principle of benefit maximization is more likely to help us in an objective way if used in conjunction with the second principle we are going to discuss. The two principles looked at together may serve us well, as they provide checks and balances against any self-serving motives that may otherwise be lurking in our decision-making.

The Principle of Equal Respect   You will again remember our discussion of this principle in Chapter 2. When we use this principle, rather than concentrating in our moral decision-making on the potential benefit of that decision to the greatest number of individuals, we concentrate much more on the appropriateness of considering all individuals as being worthy of equal treatment, simply because they are human. The humanity of individuals is the common denominator here rather than, for example, their cultural or socioeconomic status.

You can readily see that when we consider the two hypothetical situations we discussed in the section on the principle of benefit maximization, and incorporate into our thinking the principle of equal respect, it would be difficult to continue to feel justified in taking the law into our own hands. Would we wish similar treatment to be handed out to our son, mother, or grandparent? How would we feel if we were on the receiving end of such treatment? Would we wish people generally to be treated in this way?

I think we could reasonably assume that most officers would respond to these questions in a way that would indicate they are committed to fairness and justice.

So we can see here, by stopping to consider this principle before we make a decision, we may radically change our thinking about a particular course of action and our subsequent behaviour. And, equally important, the danger can be reduced of falling into the trap of rationalizing an act that could get us into deep trouble.

If we think about it for a moment, it becomes clear that the notion of due process has its roots in this principle of equal respect, as does our thinking about the basic rights of individuals in the kind of society in which we live. Due process, though a legal term, is very much a moral term also. As you know, the concept of due process has to do with developing and applying fair procedures for making important decisions affecting the lives of others. There is something ethical about such an approach. As a private citizen, I take comfort from it, even though from time to time the process can let us down.

Reflections

Why should we be aware of the inherent dangers of using only one ethical principle to justify a decision?

Think of a moral dilemma you have encountered. How would you use either or both of the ethical principles to assist you in your decision-making?

Is an accused person or an offender entitled to benefit from the principle of equal respect?

Is it possible, do you think, to make a good case for sometimes suspending due process?

The Bell, the Book, the Candle

These tools for moral decision-making have been suggested by Michael Josephson of the Josephson Ethics Institute in California. They are simple and straightforward to use, but can be a most useful check against which to measure a course of action we are considering. I would suggest here—and this is a point that you may wish to think about for yourself and also to debate in class—that if we are contemplating doing something that we feel reluctant to subject to the scrutiny of tools like these, then we probably know already whether we should take that step. Our reluctance probably indicates what we know either instinctively or intellectually that what we are proposing to do is not morally justifiable.

The **bell** simply refers to a question we might ask ourselves when we are about to make a choice: "Are any warning bells going off?" If so, then we should at least stop and reconsider what we are proposing to do. Of course, the warning bells may be going off simply because of a particular way in which we have been socialized or conditioned, where the proposed course of action is still a defensible and ethical response. Normally, however, if the warning bells go off, and we wish to avoid for ourselves much subsequent grief, we would be wise to heed their warning and respond appropriately.

The **book** is a tool that simply refers to laws, regulations, codes of ethics, and other forms of written guidance. So when I am contemplating a course of action there are a number of questions I can ask myself. Is what I am proposing to do lawful? Does it fall within the department's policies and procedures? Does it potentially violate any section of the department's code of ethics?

The **candle** simply refers to how the decision might look if it were exposed to the light of day, as it were, or to public scrutiny. If I choose to act in this way and it is examined in the public arena, can I defend my action? Will I continue to feel good about my action, and will I continue to feel justified in doing what I did?

Reflections

Is it possible to quickly review the bell, the book, and the candle in heat-of-the-moment decisions? Or are these tools only useful when considering moral conflicts, where a quick decision is unlikely to be an issue?

Have you ever consciously used the bell, the book, and the candle, or something approximating them, in your own moral decision-making?

If your answer is yes, was the exercise helpful as a guide in your decision-making?

The Imaginary Video Camera

We live in a technological age, and one of the curses or blessings of this age—depending on one's perspective—is the video camera. Many unsuspecting souls, as we have seen on the news and in documentaries, have discovered to their great cost that they are the unwitting "stars" of video footage. And included among these souls, as many of you will have seen, have been a number of law enforcement personnel.

I am, of course, talking about those embarrassing moments when inappropriate conduct is captured by some amateur sleuth in the vicinity, armed with her video camera. For example, many of you reading this will remember the graphic video footage of the Rodney King beating by police officers in Los Angeles.

In August 2003 there was an altercation in the parking lot of a doughnut shop in Toronto. A visible minority male, Jama Jama, was struck in the face by a police constable, leaving him with a missing tooth and a cut face. The police officer, whose story was supported by three accompanying officers, claimed that Jama was aggressive in resisting arrest. He wrote in his report, "Male strikes out with both arms hitting me in the upper chest. Tell male to stop resisting and attempt to take control of male." The police report states that Jama's injuries were sustained in the parking lot fracas before they arrived. He was charged with assaulting a police officer and causing a disturbance.

Would you believe that subsequently a videotape of the incident was produced! The videotape was shot by Mr Jama's friend who was sitting in a car parked close by. The video footage supports Mr Jama's claim that his only involvement in the parking lot fight was an attempt to break it up. It also shows him waiting quietly for police, with his hands lowered, as they approached. The video clearly shows Jama being punched in the face by the police officer. It also shows Jama with a full set of teeth after attempting to break up the fight, a fact that is clearly at odds with the police report.

On the news the other evening there was again graphic film footage of a police officer, having stopped an unarmed motorist for a traffic violation, inexplicably firing several shots at the driver, who later died of his injuries. The driver was unarmed, and had stopped his car, at which point the officer appeared to lose his composure. Without the benefit of the video recording, most of us would have believed the officer's subsequent story that he had been attacked first.

Some interesting events unfolded in New South Wales, Australia, in 1996 and 1997. A royal commission was appointed to look into charges of state-wide police corruption. As part of the investigation, some police cars were bugged: microscopic cameras were strategically placed in the dashboard of the cars. As a result, many illegal activities by police officers were recorded and later produced as evidence to the inquiry. One particularly unscrupulous and corrupt police sergeant, after telling a television news crew that he knew nothing of police corruption, was subsequently shown video footage of himself sitting in the driver's seat of his police car, counting out money he had just been paid by a drug dealer. The dealer was sitting in the passenger seat of the police car, as bold as brass, almost as though it were his office.

You will have heard the saying "You know it's going to be a bad day when a reporter from *60 Minutes* taps at your door!" The police sergeant's experience of watching himself as his corruption was exposed for all Australians to see must have been the Australian equivalent.

These are just a few examples of how we can get caught out by the wonders of modern technology. It would be fair to say that you and I may also at some point find ourselves on the receiving end—having our movements and behaviours captured by a video camera without our permission. And that thought alone can be a useful deterrent to our behaving unprofessionally.

Reflections

As you can readily see, this technique is an extension of the candle technique, in which we ask ourselves whether our actions can be justified in the light of day.

Is there a danger that in suggesting the imaginary video camera as a tool, I am playing on the emotion of guilt?

If you answered yes, is there a possibility that this could be construed as a manipulative move on my part?

If you think it might be manipulative, is the manipulation justified if it results in even one career being saved?

If you answered yes, why? If you answered no, why not?

(Note: You can see from these questions that this is an issue with which I wrestled prior to my making a decision to include the imaginary video camera as a tool).

What I am suggesting here is that on those occasions when you or I are beginning to feel uncomfortable, either with what we are thinking of doing, or indeed are already doing, imagining that our behaviours are being recorded may help us to gain a clearer perspective. Thinking in this way may be a quick but useful check at that moment on whether our judgment, or our behaviour, or the action we are contemplating, is sound and defensible.

Critical Thinking

Critical thinking is not so much a specific tool to help us with our moral decision-making as a general approach to problem-solving that can also help us as we struggle with our moral dilemmas.

You will no doubt be aware, through your own reading and through some of the work you have done in your classes, of the concept called *critical thinking*. There are many books on the subject, and I would encourage you to do some research and reading on the topic for yourself. In the context of this text we only have time to touch upon the skill.

You will notice that I use the word *skill* to describe the process of critical thinking. It is a skill that we can acquire through learning and discipline, but something that few of us come by naturally.

Briefly, then, what do we mean by the term **critical thinking**? First of all, the skill has little or nothing to do with being critical in the way we normally think of being critical. It doesn't mean we should go around criticizing. Rather, critical thinking means that we should refrain from taking any idea or point of view at face value—that we should examine it carefully before making a personal judgment about it. It means also that we should probably think twice about allowing others to do our thinking for us. It means that we should aim to think for ourselves, to weigh the relative merits and demerits of any idea, to be prepared to look at all sides of a given issue, and to come to our own informed opinion.

The task, however, is not an easy one. Our thinking in general, and the way we look at most issues specifically, is unclear at best, leading us often either to simply accept a point of view without examining it for ourselves, or to cling to our own—often blindly accepted or adopted—position.

In the literature, you will find many variations on the theme of critical thinking. Put simply, if we are critical thinkers,

- we will resist the temptation to jump to conclusions about issues;
- we will be prepared and able to think independently—that is, to think for ourselves;
- we will have the courage always to be prepared to subject our own views and ideas to scrutiny by ourselves and others;
- we will refrain from accepting blindly any point of view or idea that is offered, even (or especially!) if it is offered by someone we consider to be an expert;
- we can demonstrate skills of reasoning that would include the ability to honestly weigh the evidence, to trust only reliable sources of information, to disregard hearsay or unsubstantiated opinions, and to think logically about an issue.

You can see that if I discipline myself to think in this way, my response to situations involving rumour, or gossip, or peer pressure, for example, is probably going to be different from my response if I did not.

I use a handout in class that my students find helpful. It is a short piece entitled "On the Art of Thinking," by Roscoe Drummond. Unfortunately, I cannot cite the article because I am not sure how I came across it. Drummond makes the point that young men and women should be stimulated to think about knowledge, not just to acquire it.

You will recall that at the beginning of this chapter I include a journal quote from one of my students: "I have always had the notion that we never truly think while in college. All we do is memorize." Drummond writes something similar: "I was so busy learning facts and memorizing them so that I could pass examinations that I had no time to think about them." He then goes on to say something that should be taken quite seriously by all teachers, including me. "I cannot recall after graduation [that] I ever had the need or the opportunity to use a single fact I had acquired in getting what passed for an education."

In the article, Drummond offers some useful tips for those who wish to learn how to think critically. Here they are:

- Before you reach a judgment on a given issue, insist on getting all the pertinent facts—not just some of them—which bear on it.
- Read at least one newspaper and one magazine whose views are likely to run counter to your own. We need information we may not always relish, and we need to test our insights and our opinions against differing information and differing opinions.
- One way to help get a right answer is to ask the right questions. They constitute the beginnings of what can lead to a sound conclusion.
- One can hardly deem himself adequately qualified to defend his own view on a given matter unless he has equipped himself to expound the contrary view. It's a healthy growing experience. (*Author's note:* My students are always asked to take the view opposite to their own in the class debates we hold. They find the exercise a difficult one, but one that stretches their thinking. Try it sometime—you might like it!).
- All truth is rarely on one side in any discussion of temporal affairs. Sometimes the balance of fact and good sense will be fairly even. But when choices have to be made, this need not affect the firmness of your conviction.
- Don't hesitate to change your mind; it's a rewarding experience. Don't hesitate to hold to your convictions when you are persuaded you are right.
- A yearly mental housecleaning can be helpful. Remove from your mind all your convictions and prejudices, examine them honestly, and then decide which should be put back and which, if any, should be discarded.

These tips are good ones and worth keeping in mind when we wrestle with difficult issues.

Let me share something with you that I tell my students. Keep in mind that it is a personal opinion, and I am more than willing to hold it up to scrutiny. I tell my students how I believe we often develop our value systems. I explain to them that on occasion we accept ideas and values from others—normally significant others, for example, parents, teachers, sports coaches, or our church or school leaders—and we put them on as we would an item of clothing. Just as we generally only put on clothing we like, we tend to accept the values we "like." The problem is that if, indeed, there is any truth in what I am saying, many of our values cannot be said to be truly our own. We have adopted them, as it were, without holding them up to careful examination.

In essence, it is something like building our values foundation on sand. Because they are not truly our own values, they can be readily shifted, or challenged or changed. Because these adopted values are not our own, our investment in them can often be shallow, causing them to become quite vulnerable to any wind of change.

Reflections

Would you agree that the ability to think critically is a skill that must be learned?

How many of Roscoe Drummond's "tips" do you follow?

What is your view of the idea that many of our opinions, beliefs, and even values, are simply adopted by us? That is, that we just inherit them without stopping to think about them for ourselves?

The Other Person's "Moccasins"

I remember seeing somewhere the Canadian native expression "Never judge another man until you have walked two moons in his moccasins." In more modern parlance, we might say that it is important to look at every situation from the respective points of view of the various stakeholders. (Forgive my using the term *stakeholders* because I think it is one of those overused, cliché expressions!)

It does convey, however, the importance of attempting to see the other person's point of view when we are trying to make sense of a complex situation.

Of course, this is no easy task, even when time permits. When speed of response is of the essence, it is an even more difficult task. The wise officer, however, is the one who at least makes the attempt to develop a sense of how all parties involved in an issue, and its subsequent resolution, come to see the "facts" surrounding it in different ways. Here are some questions we might be wise to ask.

- What is likely to be influencing the accused person in terms of his explanation of the incident?
- What about the witnesses?
- Why is the shift sergeant interpreting the "facts" in the way she is?
- What about the crown attorney's perspective? Why are there some discrepancies in the witness accounts of what they thought they saw happen?
- What is their relationship, if any, to the accused, and is this likely colouring their perceptions of what happened?
- What would I be thinking and feeling if the roles were reversed, and I were in their shoes?
- What does my instinct tell me about this incident, and how does this relate to the facts as they unfold?
- Have I allowed myself to develop tunnel vision here?
- Am I so convinced of the "facts" as I see them that I am simply unprepared to take a step back to reconsider my reading of the events?

If we do become so convinced of our individual perception of what happened, first, we are clearly not being mindful of the perceptual perils of claiming to know for sure, and, second, we are covertly, if not overtly, claiming that we are the only individual capable of making the correct judgment!

Now I would not wish to suggest here that it is permissible or defensible to become paralysed by the fear of making a mistake. Some people may see such careful and informed decision-making as wishy-washy, or as fence-sitting. Let me disabuse you of that notion. Decisions have to be made, and often they have to be made in a hurry. But jumping to conclusions based on one officer's personal perspective is generally not an approach that will lend itself to informed and wise decision-making.

A standard check might be asking something like "How would I feel in this situation if I were this person? Am I looking at the entire situation, and am I responding to it in as objective a way as possible?"

Attempting to put ourselves in the other person's moccasins might be one step in the right direction. By so doing there may be a better chance that justice will be served.

Reflections

Can you think of any case where justice was not served because the investigating officers jumped to a premature conclusion about an accused person's guilt or innocence?

If you can think of such an incident, might the process described here of attempting to see the situation from many perspectives have helped to bring about a more appropriate and less embarrassing conclusion?

Moral Stages

One important figure in the area of moral development and moral education is Lawrence Kohlberg, mentioned earlier. I would encourage you to look up some of Kohlberg's work, and to engage your facilitator in a discussion about his ideas.

It is my hope that Kohlberg, now deceased, would forgive me for reducing what can sometimes be fairly complex theories of moral development down to what can only be described as a simplistic level. His work advocates that humans operate at any one of six stages of moral development. The stages range from 1, which is the least morally developed and principled, to 6, which is the most developed and highly principled.

If we are at stage 3, for example, the motivation for our behaviour comes from a need to conform, where relationships are critically important to us and govern our moral decision-making (ring a bell?).

Stage 4 is the stage at which we make our choices based on what Kohlberg calls "conscience maintenance." If we don't feel guilty, we must be doing all right.

Stages 5 and 6, on the other hand, are stages at what he calls the "post-conventional" level, where individuals are able to make reasoned moral judgments based on sound ideals and ethical principles, and where those individuals are more likely to have the moral courage to follow through on their moral choices.

Now what is the point of all this? (And I am sorry if this sounds convoluted!) If we think of stages 3 and 4 as basically "going with the flow," and stages 5 and 6 as principled and courageous, we may ask ourselves an important question just prior to making a moral choice: If I make the moral choice I am about to make, at what stage might I be operating? Is my choice based more on what others think of me and what I think of them, or is my choice one that I can truly call my own and one that is reasoned, principled, and strong?

It's a tough question, but it can be a useful check on our behaviour.

Reflections

At what stage do you think you generally operate in the ethical choices you make?

What do you understand from the term *moral courage*?

PUTTING THE TOOLS TO THE TEST

Let's have a go at working through a moral dilemma using the tools. We will do so briefly, but the process may take a little longer in class because of the diversity of opinion that no doubt will be brought to the issue.

As this chapter was being written, I had lunch with a chief of police, and during our chat he related a dilemma he had been thinking about. He said, "Geez, I've got this problem. We have some neighbours several doors up from us, and they are absolutely delightful people. We have known them for several years." He explained that the two families had socialized together. "Their son is a graduate from a college justice program, and desperately wants to get on with my force. But you know, I kind of have some reservations about him. There's something that I can't put my finger on, I feel kind of uneasy about him. Obviously his parents would like me to help because the kid has got his heart set on policing, but I don't know."

So, briefly, let's try to apply the tools to the problem. The dilemma is, Should the chief help his neighbour's son to get hired by his police department?

The first tool, **setting limits**, appears to have little or no direct application to this problem.

The second tool is **ethical principles**, the first of which is the principle of **benefit maximization**. Now if we apply this principle to the problem, we quickly see that the greatest benefit, if the lad were hired as a result of the chief's influence, would accrue to the son and his parents. The fact that there may be a much better candidate available means that the interests of the community probably would not be well served. Overall, therefore, it is likely that less public good would result if the son were hired. The second principle, **equal respect**, refers to giving all individuals an equal amount of consideration. That would clearly not be the case if the chief gave this lad an edge by exerting his influence. So far, then, the tools would indicate that the chief should not use his influence to help the neighbour's son.

The next tools we discussed were the **bell**, the **book**, and the **candle**. In this instance, I am unsure about what the chief, if he were to use his influence to help this lad, would say about the *bell*. Only he would know. The *book*, however, may be helpful. Police force policies would most certainly address issues of conflict of interest, the importance of fairness, and the inappropriateness of senior officers using influence for personal reasons. (This may mean, too, that warning bells would be going off.) The *candle*, referring to whether a decision will stand up to public scrutiny, would indicate that the chief should not use his influence in this case. Clearly, most members of the public would wish police hiring decisions to be based on fairness and equal opportunity. So far, so good: the chief would be unwise to use his influence.

The next tool, the **imaginary video camera**, would probably also indicate that it would be unwise of the chief to help his neighbours, because if his use of influence in this way were recorded or documented, he would probably have to answer some hard questions. The message continues to be that he should not help the lad any more than he is prepared to help any other candidate.

Critical thinking is the next tool. If the chief is prepared to put his decision-making to the test he has already submitted himself to, he has begun to engage in critical thinking. He would now begin to see the issue more clearly, and the decision-making process would be elevated from the gut level to a level where neighbourly loyalties are probably not going to dictate the decision.

When we consider the chief placing himself in the *other person's "moccasins,"* an important question to ask might be, If I were a candidate for this police force and I thought I was placed at a disadvantage because of a senior officer's "pull," how would I feel? The answer would appear to be self-evident.

Conclusion: the chief should not use influence to help the neighbour's son.

Finally, can consideration of Kohlberg's **moral stages** help us here? Well, if the chief wants to be liked, and wants to reinforce to an even greater extent his neighbourliness with the lad's parents, he can help the lad. But would this be a good reason? Perhaps it would be better to think the whole thing through independently, and then make the right decision, irrespective of any potential fallout with the neighbours. Loyalties, therefore, would not be permitted to "colour" his thinking about the situation, nor to dictate his decision.

It would appear that the chief can make his decision with confidence and move on. The agonizing is over. To use his influence in this case would be unfair, and would not withstand public scrutiny. And perhaps most important, the general good of the community would be better served by hiring the best candidate available.

My hope is that this exercise demonstrates how our moral decision-making can be helped by submitting the dilemmas we encounter to a similar process.

SUMMARY

Perhaps the acid test underlying all of these "checks" is, How do I feel about myself? Do I feel strong? Or do I feel awkward and ill at ease about my action? Do I feel shaky inside? Am I depressed about the choice I made? Do I find myself constantly looking for ways to justify my choice?

I think it would probably be safe to say that few of us, and I include myself here, have not experienced the soul-searching indicated by troubling questions like these.

The biggest worry of all, however, is the person who has *never* felt this struggle, for none of us is perfect. The individual who has experienced the struggles can, at least, take some comfort from the fact that her moral sense is still active.

We are human, and part of being human means that we will make mistakes. However, the mistakes may be fewer, and the really big, career-destroying and peace-of-mind-destroying mistakes may be fewer, if not eliminated, if we adopt some of these techniques to help us work through the moral dilemmas confronting us in our work.

These tools may help us think more clearly, and they may serve to reinforce our moral sense, but the entire issue boils down to one of moral courage. I know what I should do because I have thought about the dilemma carefully and honestly, but do I have what it takes to follow through?

Moral fortitude does not come easily or readily.

Let me leave you with one final thought in this chapter. Does the courage to act come more readily for our having systematically, honestly, and thoughtfully weighed the consequences of our decisions?

chapter nine

Where Do You Stand?

There is always more than one choice to be made. I do not think there are right and wrong choices, but rather better and not very good choices.

> —Journal entry. Justice student in ethics course.

I am a strong believer that when people choose to work in the justice field, the most important factor is honesty. You must be honest to yourself as well as others.

> —Journal entry. Justice student in ethics course.

I believe that if a person thinks in a certain way then he/she should be able to give reasons to support why he/she thinks that way.

> —Journal entry. Justice student in ethics course.

I have always found it curious that as a society we discourage people from changing their minds. How did it become a sign of weakness and not strength? For me, one of the most admirable qualities in another person is the ability to say, "I changed my mind, I never thought of that before." In my view, that indicates strength and intelligence and open-mindedness.

> —Journal entry. Justice student in ethics course.

INTRODUCTION

As you can see, this chapter has no Chapter Objectives section. The purpose of this chapter is simply to give you some material for your own thinking, and for discussion with your classmates.

You may remember my stating in Chapter 1 that the dilemmas included in this chapter are based on actual experiences. They are not hypothetical. They have been directly or indirectly recounted to me by officers in the field, so I hope they will ring a few bells for you.

The police and corrections scenarios are not separated, and this is deliberate. You are encouraged, whatever justice profession you are studying for, to have a go at working through several, if not all, of the issues. In this way you will develop some insights into the kinds of awkward problems that can confront both police and correctional officers.

When you are thinking about the dilemmas and how you or your colleagues would probably respond to them, attempt to place yourself in the actual situation. It is one thing, in the comfort of one's home or the relative comfort of a classroom, to imagine how we would respond. But I want you to try to transport yourself into the situation and imagine the dilemma unfolding. Try to grasp all the nuances of the situation: the subtle and not-so-subtle pressures, the friendships, the comradeship, the expectations you have of yourself. Think about who you are as a person. Think about the faith your parents and family probably have in you. Look at each situation in its entirety and attempt to weigh all the factors before you begin working out what you think you would do.

Let me offer a word of caution. If you read a dilemma and you think you know immediately what you would do, there is a good chance that you have not grasped the complexity of the issue. They are dilemmas *because* they are likely to induce in you a similar response to the one experienced by William Tafoya's subject: "I guess I was damned if I did, and I was damned if I didn't."

Please also be aware that thinking we know what we *should* do in a given situation, and thinking we know what we *would* do in that situation may be two different things. If there is a discrepancy like this between your thinking and your anticipated behaviour, what might account for the discrepancy? Why is it, do you think, that at times even when we know fairly certainly what we should do, we can find ourselves lacking the moral courage to follow through?

Some of the dilemmas relate to police officers, others to correctional officers, and others to both professions. Two scenarios ask you to consider a dilemma facing a senior officer, so that you can begin to realize how difficult it can be at times for administrators when they, like you, have to wrestle with what should be done.

You are not being asked, however, to tackle these dilemmas without help. I would recommend that you go back to Chapter 2 and review the discussion there. Look at what we had to say about the nature of a moral dilemma. Remind yourself of what we said about gullibility and mindlessness; about what Dr Martin Luther King, Jr had to say about good laws and bad laws. Refresh your memory on libertarianism, determinism, utilitarianism, and the principle of equal respect. Think again about what Plato, Socrates, Aristotle, Hobbes, Kant, and Sartre had to say about justice. If you learn to incorporate these concepts and ideas, you may find your thinking becoming clearer and more comprehensive.

May I also suggest that you practise using the tools discussed in Chapter 8.

Have fun with the dilemmas. But at the same time remember that one of these days you may find yourself in a similar situation for real. I hope the experience of thinking privately about them, and then discussing these dilemmas with classmates, will help you when the time comes.

You should not feel "led" in any direction by the questions. They are not designed to move you toward thinking in a prescribed way about an issue. They are included simply as prompts for thought and discussion. What matters most of all is what you *think* and *feel* about these issues, whether or not you can *defend* your position on them, and what *reasons* you employ to support your response.

Remember what I said earlier about moral courage beginning where we are willing and able to take a hard, honest look at dilemmas like these; and where we are strong enough, at least for a few minutes, to put aside our prejudices, biases, and preconceived notions.

Enjoy the experience. You may learn something important about yourself, your colleagues, your chosen profession—even life in general!

Note: Some of the following dilemmas contain graphic language that may be offensive to students. These incidents did occur, however, and are reported verbatim. It is my sincere wish that readers will understand that the language is used neither gratuitously, nor for effect; rather, it is a true reflection of the actual incidents and relates to important work-place issues.

DILEMMA 1	**Is Honesty a Personal Quality— Like Shyness?**

When I was department administrator of the justice department of the college where I work, two Police Education students were brought to my office by a faculty member. They had been caught cheating on an examination.

I asked them both if they had, in fact, cheated. A flood of excuses followed, accompanied by interesting you-go-first glances from one to the other. It was only after the cheat sheet was produced and clearly matched one of their answers that they grudgingly admitted to cheating.

Then something interesting happened. One of the students became quite emotional, admitted his guilt, said he was ashamed of himself, and said he had learned a painful lesson from the experience.

The other student, on the other hand, continued to be defiant, insisting that many students cheated on examinations. I asked him if he thought that honesty was an important personal quality for a police officer, and he replied, "Well, it would be different if I had a uniform on." There is a curious logic to this response.

Both students were asked about their career aspirations. The defiant student said he wished to join the Ontario Provincial Police force, and the other student said he wanted to join the Royal Canadian Mounted Police force.

Questions

1. Is being honest a personal quality or characteristic—like shyness, for example— or is it something we simply decide to be, depending on the circumstances in which we find ourselves?

2. Is there any moral difference between cheating on an examination and stealing from an unlocked electronics store while on night patrol as a police officer?

3. If you were me, and you heard six months later that the defiant student was in the final stages of the recruitment process for the Ontario Provincial Police, would you, or should you, pick up the phone and give the recruitment section the information you have?

4. If your answer is no, why not? If yes, why?

5. Is a college professor's or administrator's primary responsibility to his student, or to the welfare of the community?

6. If a person appears subsequently to be genuinely remorseful for a wrongdoing, should this be a mitigating factor in the way the situation is dealt with?

7. Imagine a situation at a police or corrections training centre where in, say, the tenth week of a 12-week orientation course, a student is found cheating on an examination. Would this circumstance warrant, in your view, dismissing that trainee officer from the force?

8. Which of the tools, if any, may help you with this dilemma?

DILEMMA 2	**Thank God for Sick Days**

Note: This is a dilemma that can be applied across the board in the justice field. Please be aware that this scenario has the potential to put you on the defensive. Please try to look at the situation as calmly, logically and reasonably as you can.

South Side Detention Centre is a maximum security institution housing inmates on remand or serving sentences of 90 days or less. It is a government facility, so the staff are public servants.

Recently there have been several reports in the newspapers about the troubling levels of sick leave used by public servants. The average number of sick days (or "sickies," as they say in Australia) taken during the last year by government workers has been reported and confirmed to be an average of 11 for each employee. In correctional institutions and police forces the average number is 15 per employee. The cost to the taxpayer is estimated to be tens of millions of dollars each year because of the overtime or casual staff needed to provide shift coverage.

The attitude of many of the staff at South Side Detention Centre is that the sick days are one of their benefits. They feel that sick days are provided for in the contract and are there to be used up, whether they are sick or not. "If you don't use them, you lose them," appears to be the prevailing attitude.

You are an officer at the detention centre, and one day you are sitting in the lunch room and you hear the following conversation.

Officer 1: I owe this place nothing. I make no bones about it, if I want to use one of my sick days for a day off, I use it, that's all there is about that!

Officer 2: I know I haven't been here long, I don't know too much about what goes on here. My understanding was that those sick days are there for us in case we really are sick.

Officer 1: That's the way most of us felt at one time, but not anymore. If I want a day out with my wife, or a day out fishing, I'm out of here on a sick day.

Officer 2: I'm not sure I would be able to relax. I think I would feel guilty all day long, so I may as well go to work. Besides, what if you get caught?

Officer 1: So what? The boss does it too. We all do it. It's a right. It's in the contract.

Officer 2: I don't mean to sound self-righteous or anything, but isn't it a form of stealing? If I phone in sick when I am not sick, if I go fishing on a sick day, I am being paid to go fishing, aren't I? I am not sure about that.

Officer 1: It can't be wrong because everybody does it.[1]

Questions

1. Is this problem primarily an economic one, or an ethical one?

2. Is it morally defensible to go fishing on a sick day when I am not sick, but when that activity will probably enhance my mental health?

3. Does a manager or senior officer have a greater responsibility to act morally than a frontline officer has?

4. Is Officer 2 in the dialogue responsible only for her conduct? If she becomes aware of an abuse of the sick leave policy by a colleague, is she morally "on the hook" to do something with that knowledge?

5. Some people say that silence implies consent. What do you think?

6. Let's assume for the sake of argument that a sick day is worth $200. If I take a sick day when I am not sick, isn't this something like taking $200 out of petty cash? If not, why not?

7. Which of the tools, if any, may help you with this dilemma?

8. To what extent, if any, do you think your response to this kind of dilemma would be governed by officer subcultural constraints?

---

1   In August 1999, Statistics Canada released its figures on absenteeism rates in Canadian workplaces. In 1998, the average full-time employee was absent from work an average of 7.8 days. Those with sick leave benefits were more likely to phone in sick. Unionized workers took an average of 11.4 sick days compared to non-unionized workers, who took an average of six days. The cost to employers is in the billions of dollars (*Toronto Star,* 15 August 1999, pp. A1, A10).

DILEMMA 3	The Timing Stinks!

The Mount Henry Police Force (the name is fictitious) is a small, municipal force in a semi-rural community in Canada. The force has a chief, deputy chief, five sergeants, and 37 sworn officers. Generally, the force is considered to be well run, and enjoys the respect of the community it serves. The chief is considered to be a politically active individual, and is on good terms—is friends, even—with almost all of the local officials.

It is a damp, cold November evening, and Constable James has parked his car in a rest area a few kilometres from the station to get caught up with his paperwork. James is a probationary constable, with just three months of his 12-month probation remaining. The roads are quiet, and he is enjoying a few moments of peace to collect his thoughts, and to reflect on what has been a long shift.

At that moment the police car in which he sits is illuminated by the headlights of an oncoming car. The headlights are on full and dazzle James for a moment. The car passes him slowly, and when he turns around to take a better look at it, he notices it is being driven erratically. "Thank God," he thinks to himself, "there are no other cars around."

He starts the cruiser, turns on his lights, and swings around in pursuit of the car. As he approaches the car, he notices it is increasing in speed. It veers alarmingly from side to side, and the young officer becomes increasingly anxious at the thought of what might happen if another car happens to approach from the opposite direction.

He makes a quick decision that this is simply too great a risk to take. He accelerates, catches up with the now speeding, veering car, and manages to overtake it. Getting ahead of the car, gradually he is able to slow the driver down and get him to come to a stop.

Shaken, and feeling quite angry, he gets out of his cruiser, approaches the vehicle—and shines his torch into the face of the chief of police. The chief explains that he had been to a difficult budget meeting, and afterward knocked back a few drinks. He had assured his friends, who expressed concern for his safety before he left, that he was fit to drive, especially since it was not far to his home, and along a quiet road.

"Constable James, you're a good man. My house is just up the road; I'll be fine. Good work. I'll remember you for understanding the situation. Those buggers at City Hall don't understand what these budget cuts are going to mean to us. Anyway, whatever happens, even though you are the youngest officer on the force, you have my word that you will not be affected...if you see what I mean."

Questions

1. What should Constable James do?

2. What would you do if you were in a similar situation?

3. Constable James has to think about his wife, his baby, a mortgage, and a career he loves. Should his decision be influenced by these responsibilities and concerns? Where should one's priorities lie in situations like this one?

4. Keep in mind your responses to questions 1, 2, and 3. What if you were a police officer pulled over for drunk driving, and the investigating officer, a colleague and friend of yours, chastised you for being a stupid #$*!, but said he was going to give you a break. Should you—would you—accept his kind offer?

5. Should the law be applied in one way for the police "brotherhood" and in another way for ordinary citizens?

6. Which of the tools, if any, may help you with this dilemma?

7. To what extent, if any, do you think your response to this kind of dilemma would be governed by officer subcultural constraints?

DILEMMA 4	Only Thick-Skinned Women Need Apply

The detachment for which you work is in a medium-sized city, and part of a national police force. You are a male officer and you have worked for that police force for the last three years. Generally you enjoy your job, particularly the friendship and camaraderie of your colleagues.

There are 13 officers in the detachment, one of whom is a 23-year-old woman in her first year of work as a police officer. You think she is a dedicated and competent officer, and you also feel that there is a place for such individuals in policing.

She sees you as a decent person with good values, and she also sees you as supportive of her. She comes to you one day, when you are both away from the detachment, and asks if she can speak with you in confidence. She looks upset and worried.

She tells you that in the last six months she has

- been told by one sergeant that she would not be a real woman until she had a child, and that she should go home and start a family, preferably with him as the father;
- been grabbed and kissed by a male colleague officer who told her to call him when her husband was away if she wanted a "real man";
- been subjected to viewing a pornographic movie on a night shift in a joint work area; when she complained, her colleagues told her to "hit the road";
- found a pair of plastic breasts taped to her work station with "Police Girl" written on them;
- complained to a supervisor she thought she could count on for support, who advised her to "lighten up."

What she has just told you comes as no surprise to you because these "events" have been common knowledge at the detachment. In fact, when the male officers have been together, you have joined them in having a good laugh about the situation.

She trusts you, feels an acute sense of injustice and embarrassment about what she has been subjected to, and wishes to pursue the matter by laying formal harassment complaints against the five officers involved. While you have some significant misgivings about the way she has been treated, you see these five colleagues as good officers, and you enjoy their comradeship. She asks you for your help and support. She wants you to corroborate, as far as you can, her account of the events. But more than anything, she wishes you to help her through what is going to be an extremely difficult ordeal for her.[2]

Questions

1. Should you, and would you, lend the support that is being requested of you?
2. Did you smile when reading this scenario?

3. We have all been subjected to and embarrassed by workplace humour. I remember being sent to a store when I was young to buy a left-handed screwdriver! And one of my former students, all 300 pounds of him, during his first week as a correctional officer, was told to supervise a young offender in the sick room. The shift supervisor told him the young offender was highly dangerous (he wasn't) and that the student was to wear a hockey helmet (several sizes too small) all the time he was supervising the offender. The joke terminated when one of his colleagues took a photograph, recording the event for posterity.

   Are the experiences of the female officer in the dilemma something like the situations I have just described? Or is there something different about them?

4. What are your views about women working in (formerly?) non-traditional occupations, particularly as police or correctional officers?

5. Which of the tools, if any, may help you with this dilemma?

6. To what extent, if any, do you think your response to this kind of dilemma would be governed by officer subcultural constraints?

---

2 "Female RCMP officers sexually harassed at work. Six out of 10 surveyed were affected, internal study says" (*Globe and Mail*, 27 September 1996).

DILEMMA 5	I Guess He Should Have Stayed Home

Ted Egan has worked at White Plains Correctional Centre for 13 years. He is a dedicated, competent employee. He is well respected by his colleagues and by the administrators of the institution.

Several times during the past 10 years, Egan has been approached by senior officials and encouraged to apply for promotions as they became available in the institution. Each time he declined, explaining he was happy doing what he was doing because it meant he could spend more time with the inmates.

But lately things have not been going very well for Egan. His wife of 21 years has just announced she is filing for divorce. She has met another man. In addition, Egan's 13-year-old daughter is acting rebelliously. Egan, particularly because of the business he is in, is keenly aware of a serious breakdown in communication between himself and his daughter. The marital problems are certainly not making matters any easier.

Now Egan is at work. It has been a difficult shift on this particular day. He reported for work at 7:00 a.m. after an emotionally draining evening and night dealing with problems at home. Clearly, he arrived at work tired and worried. As is often the case at stressful times like this, the inmates seem to sense he isn't his normal self, and start to act out.

One particular inmate is getting on Egan's nerves. He is 18 years old, small and "yappy." He never stops. For several hours he has been carrying on with his high-pitched, pointless, and increasingly maddening "yap."

Egan somehow manages to remain in control of himself; that is, until this particular individual comes into the unit office to complain about some of the other inmates. They had told him in no uncertain terms to "belt up or else." At this moment Egan loses control and slaps the inmate across the face. The blow is hard enough to leave a red welt and a small cut. The noise of the blow is loud enough to be heard by the other inmates—who spontaneously break into a round of applause! In 13 years, this is the first time that Egan has lost his temper in this way.

Another officer is called to cover Egan's post, and he goes to report the matter to the superintendent. She invites Egan in, and listens intently to someone she normally admires and respects as he recounts the incident.

"I did it," he said. "I hit him hard. He just got to me. Things have not been good at home lately—but that's no excuse. I know you have to fire me. Or would you let me resign? I'm truly sorry it had to happen this way."

Questions

1. If you are the superintendent, what should you do? Should you fire Egan? Should you allow him to resign? Should you perhaps take a course of action that is quite different from these two?

2. Should an officer, in a position of public trust, be fired for making one uncharacteristic mistake when there appear to be extenuating circumstances contributing to that mistake?

3. Should the superintendent, in making her decision, take into account the probable effect of Egan's dismissal on the morale of his colleagues? Or is this simply not an issue?

4. Assume that the superintendent made a decision not to fire Egan; let's say in this instance he was warned and reprimanded. Might the superintendent be in danger of signaling that it is okay to engage in unprofessional conduct, as long as it occurs only infrequently?

5. Should Egan's personal circumstances, and his subsequent attitude of contrition, be a factor in the superintendent's decision?

6. Which of the tools, if any, may help you with this dilemma?

7. To what extent, if any, do you think your response to this kind of dilemma would be governed by officer subcultural constraints?

DILEMMA 6	**A Little Bit of Intimidation**

You are a correctional officer at Cedarvale Correctional Centre for Youth. The centre accommodates young offenders on remand. Their ages range from 15 to 18.

You are working the afternoon shift from 3:00 p.m. to 11:00 p.m. and, to put it diplomatically, the shift has been "lively." In fact, to put it less diplomatically, it has been the shift from hell! You and your partner are both beginning to fray around the edges. At one point you hear your partner shout at one of the kids, "I have one nerve left and you are getting on it."

One inmate in particular has been getting under the skin of your partner. He is a bullying, loud, unkempt, and obnoxious individual. According to the morning shift log, he has been testing the limits all day, and he is still at it.

Your partner suddenly decides that he has had enough. He tells this particular young offender to come to the office, and asks you to come as a witness.

No sooner have the three of you entered the office than the inmate starts mouthing off again. It seems there is simply no stopping him. He knows no fear. He certainly is not intimidated. "I can't wait to get out of this fucking dump. I'm sick of the fucking slop you give us to eat 'round here. I'm sick of your silly-assed rules," he exclaims. "Fuck the lot of you!"

By this time your partner is getting red in the face, and is clearly close to losing control. "Now look here, you puke-faced little prick," he responds. "You've been in my face all shift long, and I've had it with your fucking attitude. Frankly, you make me sick, and if I were free to punch your fucking lights out, believe me, I would. You're smart enough to know we can't do that anymore—you punks have more rights than we have as staff. Anyway, let me tell you something. Since you are going to get out of here, and since you're heading right back to the street and all of your dangerous fucking games, I have no doubt you're going to get right back into the same shit you were in before."

By this time your partner is on a roll! It appears there is no stopping *him*. He is in full flight, all right. He continues, "But let me tell you one thing, you little shit-face, I can't wait for the day when you get to the big house. That'll wipe the smile off your face—the first time they rape you—they like fresh-faced boys like you. It'll look good on you. Now get out of my face, and remember this conversation the first time they stick one up you."

Questions

1. This inmate complained to the administration about the manner and content of your partner's "chat" with him. You are asked to write a report about the incident. What do you think you would write?

2. Is it ethical to use fear as a supervisory technique?

3. Do you think you would use a similar technique under similar circumstances?

4. Let's assume that your partner's chat did the trick—that the inmate quieted right down, and as a result became much more manageable for all the officers. Let's

also assume that you feel your partner's technique is a dubious one. If the end result is a good one, does that always justify the means by which you achieve it?

5. If an officer uses a fear-inducing technique, is her motive for using that technique important or relevant in deciding whether such a tactic is ethical?

6. You may have heard of a program called Scared Straight in which convicted felons, usually serving life sentences, try to scare young offenders out of a life of crime. Are programs such as these ethical?

7. Which of the tools, if any, may help you with this dilemma?

8. To what extent, if any, do you think your response to this dilemma would be governed by officer subcultural constraints?

DILEMMA 7	Sleeping Arrangements

You are a relatively new correctional officer at Beach Head Institution. You have worked at the institution for nine months, so there are three months remaining of your one-year probationary period. At that time, with satisfactory reports from supervisors and colleagues, you will become a classified officer—a status that gives you much more security and permanency than if you were an unclassified officer.

Beach Head Institution is a maximum security facility housing male offenders, most of them in jail for having committed violent offences.

You are working a string of seven night shifts that run from 11:00 p.m. until 7:00 a.m. This is the sixth of the seven shifts, and you are feeling exhausted. You are not finding it easy to sleep in the daytime, partly because you are not yet used to it, but mainly because your one-year-old daughter is now quite active and noisy.

On this night you are assigned to work with an officer who has worked at Beach Head for the last 17 years. At 2:00 a.m., he approaches you and tells you that he is going to take the next two hours for a sleep, and that you should go on your rounds without him. At 4:00 a.m., he says, you should wake him up, and then it will be your turn for two hours of rest. You are aware that this is strictly against the standing orders of the institution, and that being caught asleep by a senior officer is grounds for dismissal.

You explain to your partner that you feel uncomfortable with the "arrangement," and you insist, for your own safety and the general safety of the area you are patrolling, that you do the rounds together. Your partner becomes highly agitated, asks you who the fuck you think you are, waves his finger in your face, and tells you that you will do as you are told. After calming down a little, he explains to you that he has an important event to attend the following morning, and that he has been banking on getting some sleep during the shift. He reminds you that you are on probation, and that if you think you can do anything about the conversation you have just had, you may as well forget it, because he would be believed and supported over you.

Questions

1. How do you think you *would* handle a situation like this one?

2. What do you think you *should* do?

3. Is it ever ethical to tell one's supervisor about the poor performance or objectionable behaviour of a fellow officer?

4. Let's assume that during the following day you make further surreptitious enquiries about your predicament with someone you trust. You discover that the lieutenants are also in the habit of taking their turns to sleep. Now what?

5. To whom are you primarily responsible:

   Your colleague?

   The inmates?

   Your employer?

Yourself?

All, some, or none of the above?

6. Some people are of the view that if one officer knows of another officer's serious indiscretion, but chooses to stay silent, both officers share responsibility for the indiscretion. What do you think?

7. Which of the tools, if any, may help you with this dilemma?

8. To what extent, if any, do you think your response to this dilemma would be governed by officer subcultural constraints?

DILEMMA 8	Arresting Officer or Judge?

This dilemma is to be looked at from the perspective of someone who is not an officer, but may have been the victim of an officer's lack of judgment, if not wrongdoing. I will attempt to relay the situation to you by simply stating what appear to be
the facts. I have made every effort not to "launder" the information. And I have made
a conscious effort not to attempt to lead your thinking on this issue.

One evening I received a telephone call from a rather distraught young woman
(a former student of mine) informing me that the evening before she had been
involved in a serious confrontation with her boyfriend.

She had told him, she claimed, that the relationship was over and that she did not
wish to see him again. They argued for approximately four hours, and the argument
culminated in a scuffle between them. He punched out the windshield of his car, hit
her across the face with the back of his hand, and pulled her hair. She attempted to
defend herself by scratching back at him. The police were called, and he was charged
with assault. She was taken to the local hospital and treated for a bruised scalp.

The purpose of her phone call to me, in reality, was not so much to give me this information, as to ask me what the process was for getting the charges against her boyfriend
withdrawn. We had a discussion about the issue, and we finished the conversation.

A few days later I was sitting in my office when this same young woman, now
seemingly even more distraught than she had been a few days earlier, tapped on the
door and asked if she could speak with me. She was accompanied by a female friend,
who seemed to me to be a reasonable, mature, and steady kind of individual. Through
tears, the young woman recounted to me how at 2:30 a.m. that morning she had been
awakened by two police officers, one of whom was a woman. She was informed of
her rights and the charges, asked to get dressed, taken to the police cruiser, and driven
to the police station. Her former boyfriend had decided to lay charges of his own
against *her*, and this arrest in the middle of the night was the result.

During the arrest and subsequent journey to the police station, the male police
officer accused her of setting up her boyfriend in what he kept referring to as a "fatal
attraction" scenario (referring to the movie *Fatal Attraction*). He made it clear to her
that she was the cause of the problem, that her boyfriend was from a good family and
had a girlfriend to whom he was committed. (The boyfriend had told her that this
relationship had terminated four months previously.)

These events have been unfolding as I write this chapter. Just a few evenings ago,
a few days after my second conversation with the young woman, she telephoned me
to thank me for my help. She told me that she had retained a lawyer who had a conversation with the crown attorney, and that the charges against her had been dropped.

Questions

1. If you were the female colleague of the arresting male police officer, and the
   facts of the arrest were as I have described them, what, if anything, would you
   wish to say to your colleague? What do you think you *would* say?

2. Is there something inappropriate about an arresting police officer making a personal judgment about the situation, and, as in this instance, comparing the details of the situation to those he has seen in a film?

3. If you think there is something inappropriate about this approach, and the matter comes to your attention (you are now the arresting officer's supervisor), what would be your response to the officer's actions?

4. Let's assume that the senior officer who is investigating the complaint from the young woman asks you—the arresting officer's partner—for your account of the events. How do you think you would respond?

5. Is it possible that there is a gender bias at play in the way the arrest unfolded and in some of the things that were said by the arresting officer?

6. Which of the tools, if any, may help you with this dilemma?

7. To what extent, if any, do you think your response to this dilemma would be governed by officer subcultural constraints?

DILEMMA 9	A Duty of Care?

Note: This dilemma is included to give you a sense of the kind of issues police managers might wrestle with, and to get you to put on the other person's moccasins....

Late in 1997, I attended a police ethics conference in Dallas, Texas, hosted by the Southwestern Law Enforcement Ethics Institute. There were many thought-provoking sessions, but one, more than any other, stands out in my memory.

The presentation was by a police officer who had himself been deep undercover for the previous five years. Delivered without much polish or fuss, it simply asked this question: "Do police managers have a duty to take care of, and be supportive to, officers who, as a result of the peculiar and high-risk nature of their work, suffer personal damage?" The presenter used his own experience as an example of how deep undercover work can "screw up" an officer's thinking and outlook on life. "At one point in your life," he said, "you have access to unlimited sums of money provided you by the force. There is very little accounting for these large sums of money. You are leading a life where gratuitous sex, booze, and drugs are yours for the asking. And you are expected to play your role very convincingly. If you don't, you could die. And then, all of a sudden, you are pulled out, and you are expected to go back to regular duty as though none of the previous five years happened."

The officer made it quite clear to those present that he had paid an enormous personal price for his undercover work. His personal relationships had suffered, his values had been severely challenged; he had suffered all kinds of role confusion; and he was severely stressed by his experience.

Permit me to add a footnote to this. During my years of travelling around conducting seminars, particularly ones in suicide prevention and hostage survival skills, I have known officers to break down and cry about incidents they experienced in the course of their work. One officer cried over a suicide he had failed to prevent 18 years earlier; another cried about storming in the middle of the night what turned out to be the wrong house, and roughing up an innocent family; another cried about ignoring deeply troubling signs of distress in a colleague, thinking he was doing his colleague a favour, until the colleague hit a career-ending crisis; and still another cried when he recalled the details of his first fatal accident involving a four-year-old child.

Due to the nature of police and correctional officer work, individual officers can, from time to time, pay a deep personal price for the duties they may have to perform.

Questions

1. Is it primarily a practical or primarily an ethical concern that managers should respond compassionately to officers harmed psychologically or emotionally during the course of their duties?

2. Should middle managers—say, sergeants and lieutenants—be especially careful about showing too much compassion, in case the troops become "soft"?

3. Should an officer's admission of psychological or emotional stress (after a serious incident) be a factor in future promotional considerations?

4. Would submission to counselling, or to Critical Incident Debriefing, by a colleague officer cause you to consider that colleague "suspect"?

5. Would your response be different if you were talking about a female colleague officer?

6. One police officer in New South Wales, Australia, went to his sergeant and said, "Can you please take me off highway patrol for a bit, because I am feeling stressed out by what I'm seeing out there." The sergeant decided to "toughen the officer up," and sent him out to patrol the most dangerous section of the Pacific Highway, a two-lane highway running north from Sidney. Several weeks later, this officer was the first on the scene when two buses collided head-on. Almost 70 passengers were killed and the ensuing carnage was beyond imagining. Three weeks later the officer went into a meadow, took out his service revolver and blew his brains out. Should or does the sergeant carry any moral responsibility for the police officer's decision to kill himself? If yes, why? If no, why not?

DILEMMA 10	It's Rough Justice, but If It Works, What's the Problem?

You work as a correctional officer in John Street Detention Centre. There are 15 young offenders in the unit. Many of the offenders have been placed in your unit because to a lesser or greater degree they manifest mental health problems.

Even though you have worked at the facility for only two months, you feel quite comfortable. The mostly long-term staff have accepted you, include you in their conversations, and are beginning to invite you to their social activities. Other than one or two minor difficulties you have with some of the inmates, which is quite normal, you have experienced no real problems, and life is generally good. You admire the skills of the experienced staff, and so far have learned much from them about how to handle difficult situations.

It is Tuesday lunchtime. You are in the dining room and you notice eight of the inmates are sitting at one table. One of them is a very immature 18-year-old (perhaps developmentally delayed, but you are not sure) named Peter. His manners at the meal table are disgusting. In fact, his eating habits are so off-putting that after seeing him eat, you are left with no appetite. One habit in particular upsets you and the other staff: he pours mounds of ketchup onto virtually every meal, swirls it around on his plate and then slurps it into his mouth. Sometimes he uses a spoon, but as often as not he uses his hands. He then eats with his mouth open and often spits the food across the table when he's talking; burps often follow. You, the other officers, and even the other inmates—some of whose eating habits are not great either—are sickened by Peter's behaviour.

This particular lunchtime, Peter's display is *bad,* and it's clear everyone sitting at the table is feeling nauseated. A fellow officer, Tom Castle, surprises you when he gets up and tugs Peter away from the table by his shirt collar. They disappear into the kitchen, and Tom comes out with a deep-sided bowl. He orders Peter to scrape what's left of his meal into the bowl, takes Peter to the centre of the dining room, places the bowl on the floor, and tells Peter to eat his meal like a dog. "Your manners are disgusting," said Tom. "You are making us all feel sick, and if you are going to eat like a dog you may as well get down on all fours like a dog. Get down on your hands and knees—you can stay there until the bowl is licked clean."

Tom later explains to you and the other staff member present that while he was acting out of frustration, he did what he did mainly as a shock tactic, to get the message through to Peter. In this way, he hoped, Peter would be less likely to alienate those around him, which would make his stay at the detention centre less problematic in terms of his relationships with the other inmates.

Questions

1. What, if anything, *would* you do in this situation? What, if anything, *should* you do?

2. Do you think Tom acted professionally? Did he act ethically? Is it ever defensible to use somewhat dubious behaviour modification techniques to eradicate socially unacceptable behaviours?

3. Assume you were offended by Tom's approach, and that his response was over the top and indefensible. Would you have intervened on Peter's behalf?

4. What should you do, if anything, if you were the superintendent of this facility and this incident had been reported to you?

5. To which of the tools would you appeal to justify your arguments?

| DILEMMA 11 | How Much Is One (Crazy) Human Life Worth? |

Note: Providing substance for the write-up of this dilemma is an excellent article written by Inspector Richard E. Dunn of the Illinois State Police, called "Ethical Police Behavior at Roby Ridge." The article was published in Command, *an official publication of the Illinois Association of Chiefs of Police (Spring 1998, Volume 8, Number 1). Also providing background is a talk on this subject given by Terrance W. Gainer, now Executive Assistant Chief of Police, Metropolitan Police Department, Washington, DC, at a police ethics conference I attended in November 1998 (Gainer was then Chief of the Illinois State Police). Chief Gainer was command officer during the entire standoff.*

The Roby standoff took place in the rural, farming community of Roby, Illinois. This dilemma centres around the length of the Roby standoff. It was certainly the longest in the state's history, clocking in at 39 days. And it was a standoff involving only one person: Mrs Shirley Allen, a widow who had lived the life of a loner since the death of her husband about 10 years earlier.

Mrs Allen behaved erratically and was convinced that aliens were out to get her. After attempting to visit her in the fall of 1997, and after being threatened with a shotgun, members of her family asked the Illinois court that Mrs Allen be involuntarily committed for her own protection. The court agreed that there was sufficient evidence and signed an order that she be detained. Three county deputies (together with Mrs Allen's neighbour and a brother) went to the home to pick her up for an appointment with the mental health authorities. Mrs Allen responded by coming out of the house pointing a 12-gauge shotgun. The deputies called for help and the Illinois State Police tactical team responded. No fewer than 39 days later, the standoff did come to a peaceful end when Mrs Allen was apprehended, but not before the tactical team had experienced some very hairy moments!

Chief Terrance W. Gainer's talk on the episode at the police ethics conference was an enlightening one. He talked of his loneliness as command officer during the standoff, of the criticism and "heat" he and his department took constantly from the media and other sources for his department's unwillingness to "blow this old woman away," and for the enormous costs built up during the 39 days that had to be borne by the force (and the taxpayer).[3]

Questions

1. Can the subsequent dollar cost of this exercise be justified on the basis that the tactical team resolved the situation peacefully and without loss of life?

2. Is it possible to put a dollar value on a human life, no matter how mixed up or "crazy" that life may be?

3. Had you been the commander of this operation, keeping in mind the enormous media and political pressure to bring this situation to an end, what limits do you think there would have been to your patience?

4. Do you think the commander's handling of the situation would have been different if Mrs Allen had been a man?

5. Would *your* thinking be different if the standoff had involved a man?

6. I am going to stick my neck out here. Feel free to have a go at it if you feel the need! After the chief's presentation, I said the following: "On behalf of the general population, and I speak as one of only a few civilians in this room, let me thank you, as a civilian, for the value you placed on a human life in that situation." What do you think of my response?

7. Does it matter if the person is crazy? Does this somehow lessen the value of his or her life?

8. One Arizona radio commentator said of this incident, "If I had been the Illinois State Police, I would have shot the bitch." Another said, "The police are chicken. I would charge the house." What do you think of these points of view?

---

3 In the spring of 1999, an interesting inquest took place in Toronto. It involved the fatal shooting by police in 1997 of a schizophrenic street person on a parked bus. His name was Edmond Yu. Three police officers were present on the bus, one of them off-duty, and the unarmed man was shot six times and killed. He had threatened the police with a hammer he had pulled out of his coat. One of the officers carried pepper spray, which was not used. The victim was a former medical student at the University of Toronto, who had succumbed to his schizophrenic demons and had started to lead a bizarre sort of street life. The police endured relentless criticism from the public and the media for what many perceive to have been an overly hasty, and lethal, response to the problem.

| DILEMMA 12 | **Working Nights Can Screw Up Your Thinking** |

You are a female (the men will have to use their imaginations here!) frontline officer in an open custody, privately run facility. Six months ago you graduated from college with your degree in Justice Studies and you are very pleased with yourself that you were successful in landing this particular job. You enjoy this particular place of work enormously because of the way in which you have been accepted, the respect you have for your colleagues and their skills in dealing with difficult clients, and the way in which you are managed. You have been made to feel that you are important, both as a person and as an employee.

The last thing you want to do is rock the boat. But you had an experience on your last night shift that left you feeling uncomfortable. You are well aware by now that working nights can do strange things to your thinking, and particularly that your sense of humour can become heightened at about two in the morning. Situations that you would not normally find funny seem to tickle you. You have come to realize how much thinking, sometimes deep thinking, you do during the long lonely hours. It's a good thing, you think, that these shifts do not have to be worked entirely alone. To work with a partner—especially a compatible, competent partner—can be of tremendous help in getting you through.

Some of the discussions with partners on nights can be quite personal; nights have a way of inducing colleagues to discuss things they would not dream of discussing, even if they had time, during the day shifts. But nothing prepared you for the conversation that transpired on this particular night shift.

You worked that shift with one of the most experienced, and respected, of the male staff. He is in his early thirties, married, with two young children. This is what he told you: "My wife and I are foster parents, have been for about five years now. The Children's Aid Society [CAS] appears to be happy with the work we are doing. It helps us pay the mortgage and the other bills. My wife stays at home to look after our kids, so it gives her something extra to do and makes her feel as though she's contributing to the household." The conversation continued in this informative vein for a while, and then, without warning, took a bizarre turn. "You know, my wife and I have good sex," said your colleague. "And we have loud sex. It sort of freaks the foster kids out—they think it's gross. But for us it's a real turn-on knowing they are listening."

Questions

1. Keep in mind that you are a relatively new staff member at this facility. You are a female. How do you think you would respond to this surprising turn in the conversation?

2. Should you report this conversation to the Children's Aid Society? To the director of the facility? Or should you simply consider it none of your business?

3. Should the possible effect on your colleague's personal circumstances enter into your decision as to whether to report the conversation? For example, it would probably mean, at least, a loss of your colleague's contract with the CAS, and might even lead to termination of his employment at the facility.

4. If what your colleague told you is true, would the behaviour of his wife and him constitute abuse or harassment of the foster kids?

5. Might there be some modelling issues here?

6. As you struggle with what you should do, if anything, to whom are you ultimately responsible? Yourself? Your colleague (and his family), who has entrusted you with this information? The Children's Aid Society? The foster children? Your facility?

7. Which of the tools, if any, may help you in resolving this dilemma?

DILEMMA 13	A Secret Compulsion

You are a correctional officer in a minimum security facility for young offenders. You are working a night shift, and a youth is admitted. This particular resident is tall, thin—almost emaciated. You begin to feel some empathy for this awkward kid, because this is the first time he has been in custody and he is clearly scared out of his wits. He is placed in secure isolation for the rest of the night because he tells the staff he is feeling suicidal.

A few days later he is integrated into the institution population and pays a great deal of attention to you, presumably because he sensed your empathy on the night he was admitted. Over the course of a few days, news about the kid begins to emerge. You look through his psychiatric file to check for yourself what your colleagues have been talking about, and, sure enough, the file confirms that this youth does not wish to remain male; he wants to be female.

One day while doing a routine cell block search, you enter this youth's cell to find him stuffing something into his pillowcase. You tell him you saw him doing this, and he looks terribly embarrassed and panicky. You ask him to show you what is in his pillowcase, and he reaches in and brings out some women's panties. With a pleading look, he explains that he prefers them to boxer shorts. He begs you not to mention this to other staff, saying that it would save him much hassle and mocking. You tell him to finish cleaning his cell. It is now time for lunch, and you have to go and pick up one of the meal carts.

On the way down to the kitchen, the scenario plays over and over in your mind. No other officer witnessed the event. This kid is obviously very mixed up, and telling the other officers might make the kid more of an outcast than he already is. Some of the officers like to make fun of kids in front of the others. You think to yourself, *What is the big deal about underwear anyway? On the other hand, if it ever came to light that I knew about his tendency and did not report it to a senior officer, my credibility would be shot and I would look like a fool.*

Questions

1. Is this youth's appetite for wearing women's underwear, and, indeed, indulging while in the facility, any cause for concern?

2. In your decision-making about whether to say anything about this incident, how much weight, if any, would you place on the youth's embarrassment and plea for privacy?

3. Assume for a moment that you have one or two colleagues who would make fun of this kid because of his preference for women's underwear. Should this fact play a part in your decision-making?

4. Is it the responsibility of correctional institutions to try and sort out "deviant" behaviours such as this one?

5. Which of the tools, if any, may help you with this dilemma?

DILEMMA 14	What If It Were You?

Note: This dilemma is addressed to males, but clearly it would be important to include female classmates and colleagues in the discussion.

You are a new male correctional officer and you have been posted to a maximum security institution. You are working your first week of shifts at the institution, and the officers are keen to fill you in on the institution, how it works, and the dos and don'ts of life as a correctional officer. One of the officers tells you to watch what you say around staff, especially female officers. This is because there are nine female officers employed at the institution, and eight of them are married to or involved with other officers.

As the officer is telling you this interesting piece of information, the one female officer who is single walks past. The officer and two of his colleagues turn to you to tell you about her. "Hey, now," one of them says to you, "there's a real catch for you if you want an easy fuck." Another officer proceeds to tell you that she is a slut and a whore who has "fucked all the single male officers who work at the institution." The first officer then says, "If you are going to fuck her, you had better use some protection because she probably has every disease in the book." The officers continue to make rude and degrading remarks about this female colleague.

Questions

1. What would you do in a situation like this? Would you sit back and listen? Would you tell them you are not interested in this information and risk alienating them?

2. Is this female officer in any way, shape, or form a victim?

3. Would your response be different if the female officer was your mother or your sister?

4. Let's assume you subsequently discovered that much of what the officers said to you was true. Is their conversation with you now justifiable because it was based on fact? If yes, why? If no, why not?

5. Some would argue that since the female officer had slept around with several colleagues in this way, she is responsible for her loss of reputation; in other words, that the treatment she is receiving (although unknown to her) is deserved. What do you think?

6. There are new officers who would welcome being included in this kind of conversation with more experienced officers because they would take it as sign of their acceptance. What would you think?

DILEMMA 15	Different Standards

You are a residential counsellor at a wilderness facility for hard-to-manage teenage offenders. You have been in your job for three months, and you are enjoying the experience.

It has taken a while, but after a couple of months you have been invited to a social evening (party!) at the home of one of the other counsellors. These staff get-togethers are held fairly often. After all, this is a wilderness camp—the nearest town of any size is many kilometres away—and even the staff homes are isolated.

You have come to like your colleagues very much. They are highly skilled in working with the youth; they are generally warm and friendly, particularly to you; and they perform their work with the kids as if it were a vocation.

The night of the party arrives, and you drive to the home where it is being held, feeling pleased that you have been included in the group. This, you think, is a terrific way for staff to relax and have fun together. But during the evening, the drugs come out—not particularly hard ones, not particularly soft ones either. Everyone at the party proceeds to get stoned…except you.

What worries you more than anything, except perhaps the illegality of what is going on, is that these same staff members regularly hold formal discussions with the youth about the perils of doing drugs.

Questions

1. There are people who would argue that what went on in the privacy of this home should be of little or no concern to anyone else, including the police. What do you think?

2. There are people who would argue that these staff members are guilty of hypocrisy in that they preached one thing (perhaps only because they were paid to do so) and practised another. How would you respond to this point of view?

3. There are those who would argue that any person involved in a position of public trust (as these staff members are) should conduct their private lives accordingly; that is, conduct their private lives in a manner consistent with their public trust and responsibilities. What do you think of this idea?

SUMMARY

These are just a few examples of the sorts of dilemmas and problems that can be encountered. My hope is that you will be challenged by them, but that at the same time you enjoy the process of opening your mind to new and interesting angles and perspectives.

I would encourage you to generate your own scenarios, preferably drawn from your own experience, and to air them with fellow students.

In the meantime, here are some more brainteasers that you might like to ponder:

- Who should police the police?
- If a recruit officer is caught cheating on an examination during the last week of his training, should he be dismissed?
- If and when you are told by your coach officer that you should forget everything you were taught at the academy, college, or training centre, what do you think you should do with this information?
- Is there a place for a whistle-blowing officer on any police force or in any correctional institution? If yes, under what circumstances? If no, why not?
- How do you think you would respond to a colleague who appears to be racist, or sexist, or is a bully? Are his attitudes or behaviours any business of yours?
- How do you feel about homosexual officers working in the justice field?
- What place is there for officer discretion? Is it possible for officers to use discretion, and still be considered fair?
- Martin Luther King said there are good laws and bad laws; good laws should be upheld, bad laws should be disobeyed. He was advocating the view that at times there can be a place for civil disobedience. How do you respond to this idea?
- Have you ever considered how you would respond if you were assigned a duty as a police officer that conflicted with your personal values? (For example, there was a case on a large metropolitan police force in which a veteran officer was assigned to ensure the safety of a doctor and his abortion clinic; the officer refused the detail on religious grounds, and was dismissed from the force.)
- One veteran police officer once told me that he knew of no police officer who had not, on at least one occasion, perjured himself. How do you respond to this?
- Is there a place for complicity in report writing?
- Would you ever lie to save a fellow police or correctional officer?
- When you are being questioned about an incident to which you were a party, is the act of omitting important details that are known to you the same as lying?
- Should police forces and correctional institutions be open to public scrutiny? When asked by one researcher for access to his officers, one police chief responded, "The (name deleted) force has not enjoyed a particularly productive experience with sociological studies of various kinds." Should there be a place for independent, unbiased inquiry into the role of policing and corrections in our society?
- Should police officers who have been fired from their jobs for discreditable conduct be permitted to go into the private security industry?

- Is it fair to have an accused person's alibi investigated by officers directly involved in the investigation of the suspect?
- Should the testimony of jailhouse informants (who normally benefit personally from their willingness to give evidence) be permitted in court cases involving serious charges?
- If you were a police officer, would you, at a friend's request, obtain confidential information on his former lover from the police computer?
- Should inmates who are mentally unwell be incarcerated in jails for criminal offences committed while they were sick?

chapter ten

Reputable Officers

You see, what's happened is that now I've learned this new material [in the ethics course] I've become more accountable. It's pretty frightening when you pay $112 [course fee] and become more accountable. Generally you get paid for being accountable. There is obviously a certain responsibility that comes with an education—this one in particular.

—Journal entry. Justice student in ethics course.

Hopefully, as we go through life we will continue to grow and mature.

—Journal entry. Justice student in ethics course.

Perhaps this meditation [the ethics course] will make us realize that much of what we hold near and dear is not valid. What if I look deep within myself and find that I don't like what I see?

—Journal entry. Justice student in ethics course.

I believe that if I go to bed tonight knowing something that I didn't know this morning, I am a better person.

—Journal entry. Justice student in ethics course.

To be part of the problem or part of the solution—that is the big question.

—Anon

THE POINT OF THE EXERCISE

I am going to assume that you have courageously had a go at some of the hard questions contained in this book; and that, either on your own or with the gentle or not-so-gentle prodding of your facilitator, you have been as honest as you can be. You may well now be asking the questions, What has it all been for? What's the point? Has the exercise been worth it?

In the final analysis, of course, you must answer these questions for yourself and come to your own conclusions. But in the meantime, let me suggest a few things to you.

First, perhaps it's better to think about the hard questions and tough issues raised in this book in the relative comfort and safety of a classroom, rather than when you are out on the street or on the institution range. I hope there will be some transferable value to the thinking and discussing you have been doing during this learning process, so that the difficult decisions you will no doubt have to make on the job will be a little clearer, a little crisper, and, perhaps most important, decisions with which you will feel comfortable. Even so, the feelings you will be left with after making these decisions, as I have suggested several times, may not offer complete peace, because the nature of a dilemma often means that there is no one perfect solution. More often than not, what we would be seeking to make would be the *best* decision rather than the *right* one. The decision you make may simply be the lesser of two evils; that is, it will still have a downside.

Second, if engaging in this exercise helps just one of you to make a decision that you might not otherwise have made, or helps to prevent you from making one that you might have made and regretted, then I would like to think my work in writing this book—and your effort and diligence in working through it—is justified. I'm thinking here of really important decisions—ones that will define you as a person and will either contribute to justice, fairness, and decency, or detract from these values that we hold in such high esteem.

But whatever your view of these two points, don't relax just yet. We are not quite finished with this process! To conclude the book, there are a few additional points of discussion in which I would like to engage you.

At this point, I am going to express some personal, and perhaps controversial, thoughts on two questions. I express these views simply to encourage you to engage in further dialogue about them, and to leave you with some final ideas to think about. I claim no special wisdom in this regard, and there are no easy answers to these issues. But I am going to take a risk and be a little more forthcoming here than I have been up to this point. I am going to lead with my chin, as it were, and set myself up on a pedestal. Of course, you are quite free to knock me off my perch if you can—as long as you can defend your views. Try to do so, but remember to consider the tools we discussed in Chapter 8. Check to see if they can support your arguments.

CAN AN INDEPENDENT THINKER BE A TEAM PLAYER?

Here's one question that you may have been asking yourself—or perhaps the topic has come up in class—"If I seek to be my own person as a police officer or correctional officer, how is this going to affect my ability to operate as a team player? Isn't being able to be a team player critical to the role of a police or correctional officer?"

This is a good question, and it is one that needs to be asked. You are correct if you think that this book essentially is about the individual officer, and the ability of that individual to make independent moral decisions and to have the courage to act on those decisions. You are also correct if you think that there may be some implications for the officer group.

Let me suggest that there is something deeply troubling, and potentially dangerous, about individuals who are prepared to give up their sense of self and their individuality simply to ensure they fit into a particular group. Let me further suggest that if any group expects of any one of us that sort of mindless, unthinking obedience, then surely we should seriously question whether we wish to continue our association with that group.

Now of course I am not suggesting that a correctional officer or police officer group is normally characterized by such an expectation. I am suggesting, however, that there may be moments when we feel that that expectation is very much there. And it is at those moments, as rare as they may be, that how we respond is critically important to our welfare and well-being.

So what are we to do with those odd moments when we feel group pressure bearing down on us to participate in an activity with which we feel considerable discomfort? Is it possible at those moments to be our own person and still continue to be accepted and respected as a team player?

I think it is. And I think so because I have been privileged to meet several officers who have been able to keep that important distinction in their professional lives. The ideas I am going to suggest here, I have learned from them.

First, I think we owe it to each other as members of a colleague group to be up front with each other. When I ask people in my seminars what names they have for officers who break ranks and report a colleague, the word "snitch" is one of the first ones mentioned. "Snitching" is a pejorative word because it implies our running behind a person's back to tell something to someone, normally a person in authority. It implies acting in an underhanded way. Often this can be a destructive, demoralizing behaviour, and I would question its value or propriety in the context of policing or corrections. Something similar can be said, of course, of other professions. But in the context of policing and corrections, where being there for each other is so important, and where trust is critical, it is imperative that we seek ways to let our colleagues know exactly how we are responding personally when something is bothering us. This will require moral courage. It will be risky. But is this approach not better than choosing to be devious and underhanded in the way we address what we consider to be a wrongdoing?

Of course, there may be times when we are so troubled by something that is happening, and so terribly scared of the possibility of colleague retribution, that the only solution is to go and speak with a senior officer in confidence. That is an option that each of us can consider when we feel we are caught.

How the senior officer responds to your request for anonymity will be interesting. If she guarantees you anonymity, and then you deliver some information about which she simply must do something, this will present a rather obvious dilemma for that officer. There's the commitment made to you by her, versus her professional and ethical responsibility to do something with the new piece of information. If the manager feels she cannot guarantee you anonymity, and she tells you this at the outset, then, of course, she runs the risk of not finding out something that is clearly important.

What would you do in this situation if you were the manager? Would you commit to holding the information and the person conveying it in complete confidence, prior to knowing what the information is? Let's continue our discussion.

What we need to aim for, I would suggest, is a climate in a policing or corrections context where we, as officers, give each other permission to approach each other in an open and honest way when something is not sitting well. If you feel this is a kind of pipe dream, something that is completely unrealistic, then we need to ask ourselves the question, Why would it be so unrealistic to expect this kind of agreement among a group of people who have sworn to uphold law and order?

What about those situations, you may ask, where we feel that simply discussing the situation with our colleague is not going to be enough. The situation is serious enough that we feel the truth must come out. Well, here I can say to my colleague something like this: "I feel really bad about your putting your uniform on when you're not on shift in order to get free meals from the Chinese restaurant, and I think it is wrong of you to do that; we're not supposed to take gratuities when we are on shift, let alone do what you are doing. It's up to you, I have spoken with you several times about this, and you have not taken me seriously. Either you come clean about it, or you will leave me no alternative but to report it. You are making all of us look bad. I can't stay quiet about this."

The officer in question, you will have noted, is not only compromising himself and his force by accepting gratuities, but gaining the gratuities by deceitful means. There are two wrongs here.

I told you I was going to lead with my chin, didn't I? Would it be a fair assumption that this suggestion is eliciting a strong emotional response in you? *No*? That's interesting. *Yes*? Why do you think you are reacting as you are?

If you are reacting negatively to what I have suggested, perhaps you can tell me how else we should approach issues of wrongdoing. If I remain silent about my colleague's inappropriate conduct, I am, in effect, condoning it. Some folk have suggested that silence implies consent, and that if I know of a wrongdoing and keep quiet about it I am equally guilty.

What obviously gets in the way of justice being served at these moments is my feelings for, and my loyalty to, my colleagues, right? These are noble sentiments, and such loyalty and concern for each other in our dangerous workplace is important. But what if it leads to abuse, corruption, cover-up, injustice, and betrayal of public trust? What then? What about the public you serve and the trust they have placed in you as a peace officer to do the right thing? What about your own commitment to honesty, impartiality, fairness, justice, equity, and human rights?

Another important question may be, Is there a possibility that if educators introduce you, the student, to this ethics experience, and it serves to reinforce your moral sense and contributes to your becoming keener than ever to fulfill your duties in a highly principled way, it might set you up for disappointment and bewilderment once you join the ranks of the officer group?

As I was first writing this book in mid-1997, there had recently been an inquiry into police corruption in New South Wales, Australia. As a result of what can only be described as a scathing indictment of the force by a royal commission, new police recruits in that state now are required in their training to participate in an ethics education experience.

In May 1997, the first 219 police graduates of the new training program were sworn in as probationary constables. "Class 268 had learnt about ethics and the 'highest possible standards' during their time at the academy, and they 'must never tolerate, never ignore' corruption, the abuse of police powers or poor service delivery," said Police Commissioner Peter Ryan. One graduate, in an interview, is reported to have said, "Just do the right thing and keep my nose clean—and whatever you see, you have to report it." Another graduate said, "I think those [corrupting] temptations are always there, but what they're trying within the recruiting system now is basically to be able to avoid that temptation, and look at moral strengths among their recruits rather than selecting people that may be swayed" (*Sydney Morning Herald*, 17 May 1997). Is this idealism, as laudable as it is, inherently risky or even dangerous for the new recruit? The group, as we have suggested, can be set in its ways, and is often characterized by the norm that the officer's commitment to his colleague group should be honoured above all else.

This question as to whether we may be setting up recruits for failure has no easy answer. We clearly do not wish you, as new officers, to leave your educational experience like all-knowing white knights riding chargers, acting as though you have a lock on morality. The officer group has its own ways of dealing with new officers who come on in this way, and the correcting experience can be a painful one!

Let me suggest, however, that any ethics education experience within the context of a justice program would be incomplete without talking about the officer group and its dynamics, including the subcultural constraints issue. I remember one graduate of a two-year police education program telling me that there had been no mention at all during her studies of what she might anticipate in terms of colleague group pressures.

Perhaps it is only by fully informing you about the subculture and how it may impact on you—the new officer or would-be officer—that you will be prepared for what to expect on the job. Then what you do with your moral idealism is up to you. Somehow you will have to quietly and humbly convey your personal standards in a way that is not going to get you on the wrong side of your colleagues.

Look around you, see who you admire. Who best exemplifies true professionalism and decency in the way he does the job? Model yourself on that officer. Tell him of your respect, and seek his advice and help.

SHOULD WE ADMIT TO WRONGDOING?

What do you think is the best policy: to admit to a wrongdoing one has committed, or not? I guess for almost every person who chooses not to admit to a wrongdoing and is subsequently found out, there are an equal number of people who benefit from covering up their wrongdoing. I have no statistics to support this view, and I may be wrong, but that's my sense of the way things are.

If I am correct in my assumption, then there would appear, at first blush, to be some advantage in at least attempting to cover up a wrongdoing, right? This would apply especially in cases where it is known that it is going to be difficult for the investigators to produce evidence of the wrongdoing.

Many of us seem to have a natural tendency to deny responsibility when we are confronted about a wrongdoing; perhaps this kind of defensiveness has its roots in our instinct for survival. There are probably few of us who have not engaged in lying in an attempt to save our skins. I am as guilty as anyone else in this regard. Would you agree that these moments when we engage in this kind of defensiveness and deceitfulness are moments that are not easily forgotten?

There are moments in each of our lives that I call defining moments. They are made up of experiences that deeply affect us in one way or another, for good or ill, and that leave us changed in some way.

Let me be quite personal with you at this point as I tell you about one such defining moment in my life. I grew up in rural Wales. I lived in a tiny village called Little Honeyborough. The village consisted of 10 cottages, two farms and a chapel. As a child I was encouraged (no, told!) by my parents to go to the chapel every Sunday. There wasn't much of a protest from my brothers and sisters and me, because when I think about it now, the chapel was the social centre of the village. (I can imagine you muttering as you read this, *Get a life!*)

One big event of the year was the Sunday School fete. It was a fund-raising, fun event, with various market stalls and innocent games booths. One year—I think I must have been about 10 years of age at the time—it was my job, along with another Sunday School "scholar," to operate a booth where the idea was to throw table-tennis balls into goldfish bowls, much as you would see at some fairs now. For threepence (a few cents) a go you could have three shots at winning a goldfish!

During the day, my fellow helper excused himself for a bit, and I was left on my own to operate the stall. It was at this point that the defining moment took place. I helped myself to (stole) half a crown (two shillings and sixpence), worth about 50 or 60 cents now, but worth much more then. The fact that the cash tin was short was noticed by my colleague when he returned. He spoke with me first, and then to one of the organizers, about the missing money. I was questioned regarding the missing money, and of course I denied knowing anything about it.

The guilty feelings associated with that incident remain with me to this day, many years later. This was a really defining moment in my life even at that early age, because it forced me to face up to the struggle between right and wrong. It also introduced me, perhaps in a way that hadn't happened to the same extent prior to that experience, to the anguish of deep guilt. In retrospect, I regretted then, and regret now, not admitting to taking the money. I often wonder how this incident might have affected me differently had I done so. In any event, there is no doubt in my mind now, that the non-verbal signals I must have been giving off during the discussions about the missing money would have literally screamed that I was the guilty party. So the organizers would have known not only that I took the money, but that I was a liar.

It's interesting, but even though I think I learned much from what may appear to others to be a rather innocuous experience, the temptation is still very much there to lie or to be evasive when confronted. It is almost as though the defensive reaction is a primeval one. My wife would no doubt confirm this for you! Am I alone in exaggerating traffic problems when late for dinner? Or in coming up with some ingenious "reason" (and it would have to be ingenious) why an important occasion has been overlooked?

Now let's attempt to relate these meandering thoughts to life as a police or correctional officer.

If I make a mistake—if I screw up, to use the vernacular—should I admit to my wrongdoing? In the final analysis, of course, we have to make our own decisions. No one should or can make them for us.

I pull the trigger too hastily in my panic; I fall asleep on night shift, and due to my neglect an inmate takes the opportunity to hang himself; I allow myself to develop tunnel vision in a case I am investigating, and because of my investment in the case I refuse to consider new evidence that indicates the accused is the wrong person; I allow myself to be duped into taking a "compassionate" letter out of the institution for an inmate, the contents of which subsequently contribute to drugs being brought into the facility.

During the investigations that will inevitably take place into incidents like these, should I be honest and above-board, or should I attempt to get myself out of the tough spot by lying?

I heard an interesting discussion once, about whether or not we should own up when we have made a mistake. A senior police officer described an incident on his force, in which one of his officers shot and killed a young male suspect. The officer had been accompanied by two colleagues. The case went to trial, and the officer was acquitted of the shooting. A few months later he retired. Several weeks after that he came forward to admit that he had lied, as had his two colleagues. Apparently the officer could not live with his guilt. Then the senior officer who was recounting the story said, "I think he is going to be okay; at least he is going to be better off than the two colleagues who lied and did not own up to their lies. They will certainly be going to jail."

Whether or not to own up to wrongdoing is a difficult question, and one you must address. Do some thinking about it now, so that you will be better prepared if and when the time comes.

Owning Responsibility

Meanwhile, let me offer you some thoughts on this question. Again, these ideas are offered simply in the interest of sparking some of your own thinking and discussion.

Would you agree that we are, generally, much more ready to forgive someone who has made a mistake and who admits to that mistake, than we are to forgive someone whom we think is clearly guilty and whom we feel is lying and cheating? This is a rather curious response, if I am right in thinking that most of us instinctively, when we have done something wrong, get defensive. Maybe this is the reason we think it is so laudable when some individuals can overcome the natural urge to lie, and be "big" enough to accept responsibility for their actions.

One of my students recounted an event she remembered from when she was about seven years old. It is significant, I think, that this young woman still had a clear recollection of the incident some 15 years later. What is also interesting is that she felt much guilt for her wrongdoing, and much relief subsequently for owning up to it.

> My sister and I were sitting at the table eating lunch, and I didn't want to drink my milk. Instead of telling my mom I didn't want it, I tipped it over and spilled it and told my mom that my sister had done it on purpose. I guess I was jealous and wanted the attention, I don't remember. I do remember the way I felt when my mom started yelling at my sister and threw her into the big brown chair. I had the sickest feeling in my stomach and I felt so bad about what I had done that I went over to the "quiet chair" with my sister (I guess to punish myself) and cried right along with her. I then called my mom over to tell the truth. I felt so relieved after doing the right thing.

And her mom forgave her—presumably for owning up to her wrongdoing.

As human beings, we are all too familiar with our predilection for making mistakes, and with our tendency from time to time to engage in wrongdoing. But even when the wrongdoing is serious, if the person making the mistake is courageous enough to admit to it, we can be forgiving. This is probably because we identify with the person as one imperfect human being to another.

You may remember one of the dilemmas we discussed (No. 5), in which an officer had uncharacteristically hit a young offender and was subsequently contrite about his indiscretion. Didn't you feel for the officer when you were discussing the dilemma? And what about the two students who were brought to me for cheating on an examination (No. 1)? One student, you may remember, was very defensive about the whole affair, and the other was ashamed and embarrassed about what he had done. Doesn't a person's subsequent response to his mistake make a difference to the way we react to him?

However, a false act of "contrition," an apology or an acceptance of responsibility for a wrongdoing that is offered without sincerity, is, without putting too fine a point on it, manipulative. As humans, we generally have our antennae out for this kind of manipulative behaviour, and it often makes us much *less* forgiving of the wrongdoer. Would you agree that we seem to have a sixth sense about whether someone is truly contrite or not? Of course, there will always be occasions when we can be duped, but normally we seem to know instinctively if a person's response is genuine.

A couple of years ago there was a heavyweight boxing match between Mike Tyson and Evander Holyfield. Some of you may remember that during the fight Tyson bit both of Holyfield's ears, tearing a small piece out of one of them. The fight was stopped by the referee, who disqualified Tyson. Later at a press conference, Tyson read what appeared to be a carefully scripted statement saying he was sorry for what he had done. But in my view, there was no *heart* in what he said. There was little of him as a human being in his words— probably because the words he read were not his own. It should not be surprising that many folk have since raised an eyebrow over this act of "contrition," and remain unconvinced that Tyson really felt that he had done anything wrong.

While we are discussing sports (am I displaying my bias here?), let's consider an incident at the Pan-American games held in Winnipeg, Canada, in August 1999. A Canadian athlete was disqualified for using no fewer than three banned substances. The Canadian roller hockey team, of which he was a member, had just won a gold medal. They were stripped of it as a result of one athlete's indiscretion. As you can imagine, this player's colleagues were not too happy with him. Compounding his transgression, however, was the "apology" this individual offered at a press conference called a few days after his disqualification and after he came out of hiding. The apology was a grudging one—an apology but not an apology. In a newspaper account entitled "Vezina Teammates Angrier After 'Apology,'" one of his teammates was quoted as saying, "I think right now I am more angry than I was before.... He basically claims stupidity.... He's claiming he didn't know what drugs were illegal. That's a crock. We knew. We had our own orientation video. As far as I'm concerned he's an absolute liar. He's not just lying to himself, he's lying to the whole country, and that's a shame" (*Toronto Star,* 4 August 1999, p. E7).

So would it be fair to say that as humans we do not respond well to people who attempt to lie their way through an investigation of wrongdoing? And that we do not respond well to people who, having been found to have committed a wrongdoing, indulge in what appears to be a fake act of contrition?

A few years ago a female first lieutenant in the US military faced a court martial on a charge of having committed adultery while serving as an officer. Lieut. Kelly Flinn,

America's first woman B-52 bomber pilot, received what is called a "general discharge." This book is not the place to debate the US military's policies around officers' sexual conduct, but it is interesting to note that military senior personnel cited Lieutenant Flinn's alleged initial dishonesty and denial as one of the mitigating factors in their decision to discharge her from the service. That would seem to imply that had there been a more straightforward, honest initial response from her, the decision might have been a different one.

A Human Response to Human Weakness

I can't tell you what you should do in those moments when you find yourself on the hot seat, just as you can't tell me. But perhaps it would be worth keeping in mind how positively we generally respond to people who accept responsibility for their mistakes, and how negatively we generally respond to people who are evasive.

There is one additional factor to think about. We should consider how we would feel about ourselves if we managed to get away with a wrongdoing by lying about it. If we are able to take comfort from, or gloat about, such a hollow victory, perhaps we should think seriously about whether or not we continue to be worthy of the important law-keeping role in society that we enjoy or are seeking.

One officer I once spoke with, who had been reprimanded for committing a serious indiscretion about which he had responded honestly, said, "At least I can live with myself; I knew what I did was wrong, but I felt that I had to take responsibility for it." When this person said this, I have to say, he went up in my estimation for the mature way in which he accepted ownership of his actions. Although the act committed clearly would meet with our disapproval, would you not agree that this officer's response is one to be admired?

Let's draw to a conclusion these thoughts on whether it is wise to admit to wrongdoing. Today as I write these notes, the report of the inquiry commission investigating accusations of wrongdoing by the Canadian military in Somalia (referred to in Chapter 1) was published. Here are some comments made by the commissioners about two of the principal players in that affair. Of Gen. Jean Boyle, the commissioners report, "We are satisfied...that Gen. Boyle was a party to the decision to informally release altered documents...and that he knew of such alterations. We are satisfied...that Gen. Boyle participated in the devising of a process which provided the public with misleading or incomplete information and condoned such a process." But of Maj.-Gen. Lewis Mackenzie, they had this to say: "Gen. Mackenzie is unique among the leaders who appeared before us.... [He] testified before us in an honest and straightforward manner. He alone seemed to understand the necessity to acknowledge error and account for his personal shortcomings.... His comportment and demeanour throughout his testimony before us were consistent with the highest standards of military duty and responsibility" (*Globe and Mail*, 3 July 1997, p. A4).

What is important is that the commissioners did not find Major-General Mackenzie blameless. He had made serious mistakes. The commission found that "Maj.-Gen. Mackenzie's fundamental failing was that he exercised inappropriate control and provided inadequate supervision" (*Globe and Mail*, 3 July 1997, p. A4). He made his mistakes, but what separated him from the other leaders in the minds of the commissioners was his willingness to accept responsibility for those mistakes.

An interesting court case unfolded in Hamilton, Ontario in May, 2001. An accused individual, Jonathan Wood, had more than twice the legal limit of alcohol in his blood on 19 June 1998 when he drove the wrong way down a highway off-ramp and killed three people travelling in an oncoming car. The case dragged on for three years because Wood appeared to be deliberately attempting to stall the proceedings. He fired four lawyers along the way and at other times excused himself from the court proceedings, citing discomfort and pain from the injuries he suffered in the crash. His unwillingness to cooperate, and his apparent lack of remorse, added to the distress of the families of the deceased who wished closure of the case. Three years after the crash, Mr Wood was convicted by a jury of three counts of criminal negligence causing death, three counts of impaired driving causing death, criminal negligence causing bodily harm (a fourth victim in the other car was seriously injured), impaired driving causing bodily harm, and exceeding the blood-alcohol limit.

He was sentenced to serve nine years and was banned from driving for 15 years. In delivering his judgment, the judge explained that part of his reason for imposing the heavy sentence was the accused's apparent lack of contrition. Said the judge, "Mr Wood is not inarticulate. He's a well-spoken gentleman who could say he was sorry. But he didn't."

Everyone in the court thought that the judge's statement was the end of the proceedings. But it wasn't. The now convicted man stood up and produced a crumpled piece of paper in his shackled hands. For the first time in the trial he looked at the family members of the people he had killed and said he wanted to address them. He said that what he wanted to say was for the families, who by now were straining to hear what he wished to say.

> I've chosen to give my statement after the judgement because it was the only way I could show you I meant what I was saying.... Despite what has been said in prior hearings...it has never been my intention to deny the accident took place nor my part in the horrific and tragic loss of lives. I have no right to request your understanding, nor do I have the right to ask for your forgiveness. But I do ask that someday you search your heart and find that I did not intend for this to happen. I would surely surrender my life if this could return your brother, your father, your husband, your friend, your son. I'm so very sorry for the grief and sorrow I have caused your family and friends. I am truly ashamed of my actions, and all I have to offer are these words, I'm sorry.

One young man in the court leaned forward during the ensuing silence. "Thank you," he said. (*Toronto Star,* 29 May 2001, A1).

Perhaps the cynical among us might think that this was a manipulative response by Mr Wood. Others might take the view that his behaviour over the preceding three years was meant to induce the court to impose the most severe sentence possible. In any event, this situation is one that gives us much cause for contemplation. One question we might ask ourselves is, Does the subsequent contrition and full acceptance of responsibility for this criminal act cause us to have greater or lesser respect for him?

I offer these examples as support for the personal opinion I have been expressing: we can be forgiving of the wrongdoings and mistakes of others. This is especially the case if we feel that they are willing and prepared to respond honestly, and to take personal responsibility for their mistakes.

Please do with this discussion what you will. Think about how you respond to people you have known who have been big enough to admit a mistake or a wrongdoing. Think about those individuals you have judged to be guilty who continue to deny responsibility. Think about manipulative "contrition."

I would encourage you to incorporate that thinking into your own personal decision-making as you do the important work you do.

A MATTER OF PUBLIC TRUST

If you are reading these final paragraphs, you have done some of the hard thinking demanded of you by this text. You are to be commended for accepting the challenge.

Perhaps your experience, like that of many of my students, has been that the process of seeking to answer the questions posed left you with a headache—that there were classes and group discussions that you left with your head spinning. Good! Your brain is a muscle just like your other muscles, and if your head ached it probably means that your brain had a bit of a workout. Or perhaps your headaches were caused by the tension you experienced because your values were challenged. I am pleased you were courageous enough to complete the exercise.

It is my sincere hope that you have benefited from this process in some small way, and that it has contributed to your development as a human being. It is also my sincere wish that you do well in your career as an officer—the public you serve is depending on you. You will enjoy—as a relatively new officer or when you are hired—an enormous amount of public trust. Keep that trust sacred, and always seek to make the best decisions you can.

You will have noticed that I used the word *best*. I could have used the word *right*. If you know what the right thing to do is, and you do it, you are to be congratulated. But you will be confronted by other issues and dilemmas that will not be quite so clear-cut. Here you will have to do what you can to make the best decision under the circumstances. At those moments, these exercises may be of some help to you.

Be a good person. Be a good officer. Always do what you consider to be the right thing, or the best thing under the circumstances, as the case may be. Uphold the law, and what you consider to be the truth; seek to carefully and honestly consider where the truth lies. Be consistent. Be fair. Beware of colleagues who may seem to have lost their moral sense, and who aim to drag you down with them. Apply the law fairly and impartially, without fear or favour.

Above all, in whatever you do and wherever you go, seek at all times to be your own person. Work out for yourself the place that colleague loyalty shall have in the way you do your job, and what the limits of that loyalty are for you.

We have two black-and-tan basset hounds at home. We adopted them in a moment of weakness and have loved them ever since! Suzie is the brains of the outfit, and Sam is as thick as two short planks. But both are lovable. One of the selling features brought to our attention by the previous owner was that Sam "sings." One mention of that word and he develops a funny look on his face, contracts his stomach and, 10 or 20 seconds later, emits a sound that would be enough to make even the most devout person doubt the existence of God. The interesting thing is that when Sam starts to "sing," Suzie starts to bark. The noise can be deafening. Our neighbours also have two dogs, and when ours bark, theirs bark. And if there were other dogs in the neighbourhood, I have no doubt they would be at it, too. Why am I telling you all this?

Here's one last thought for you to take with you. Think about it, and work out for yourself what the message may be for us as human beings. There is an old Chinese proverb that goes something like this: One dog barks at someting, and a hundred others bark at the sound.

As human beings, and as officers, are we that much different from our canine friends? You be the judge.

References

Baker, Mark (1985). *Cops*. Simon and Schuster Inc., New York.

Baron, Jonathan (1990). "Thinking about Consequences." *Journal of Moral Education*, Vol. 19, No. 2, 77–87.

Beck, Clive (1991a). "Approaches to Values Education." *Course Material: Learning Values in Adulthood*, Ontario Institute for Studies in Education, Toronto.

Beck, Clive (1991b). "How Adults Learn Values." *Course Material: Learning Values in Adulthood*, Ontario Institute for Studies in Education, Toronto.

Beck, Clive (1991c). "Moral Values." *Course Material: Learning Values in Adulthood*, Ontario Institute for Studies in Education, Toronto.

Bentham, Jeremy (1789). *Introduction to the Principles of Morals and Legislation*, reprinted by Hafner Press, 1948.

Bentham, Jeremy (1789). *Utilitarianism*, reprinted by Bobbs-Merrill, 1957.

Bittner, E. (1980). *The Function of Police in Modern Society*. Oelegeschalger, Gunn, and Haire, Cambridge, Massachusetts.

Bok, Derek (1988). "Can Higher Education Foster Higher Morals?" *Business and Society Review*, Volume 66, 4–12.

Buscaglia, Leo (1978). *Love*. Fawcett-Crest, New York.

Callaghan, Daniel and Sissela Bok, eds. (1980). *Ethics Teaching in Higher Education*. Hastings Centre on Ethics, Plenum Press, New York and London.

Cartwright, Darwin and Alvin Zander, eds. (1968). *Group Dynamics: Research and Theory*. Harper and Row, Publishers, New York.

Chidley, Joe (1995, January 30). "Bonding and Brutality." *Maclean's*, p. 30.

Christian, James L. (1990). *Philosophy: An Introduction to the Art of Wondering* (Fifth Edition). Holt, Rinehart and Winston.

Coffey, Alan, Edward Eldefonso and Walter Hartinger (1982). *Human Relations: Law Enforcement in a Changing Community* (Third Edition). Prentice Hall, New Jersey.

Collins English Dictionary (Third Edition). 1991. Harper Collins Publishers, London.

Davis, Michael (1990). "Who Can Teach Workplace Ethics?" *Teaching Philosophy*. Miami University, Oxford, Ohio, 21–38.

Desroches, Frederick J. (1986). "The Occupational Subculture of the Police." In: Brian K. Cryderman and Chris N. O'Toole, eds. *Police, Race and Ethnicity: A Guide for Law Enforcement Officers*. Butterworths, Toronto, 39–51.

Dimanno, Rosie (1995, January 30). Attacks on Kerr a punishment for breaking ranks. *The Toronto Star*, p. A6.

Folse, Kimberly A. (1991). "Ethics and the Profession: Graduate Student Training." *Teaching Sociology*, Vol. 19, 344–350.

Fox, Vernon (1983). *Correctional Institutions*. Prentice Hall, New Jersey.

Goldstein, H. (1990). *Problem-Oriented Policing*. McGraw-Hill, New York.

Grassian, Victor (1992). *Moral Reasoning* (Second Edition). Prentice Hall, New York.

Greenspan, Edward L. (2003, July 28). In defense of freedom. *Maclean's*, pp. 30–31.

Grossi, Elizabeth L. and Bruce L. Berg (1991). "Stress and Job Dissatisfaction among Correctional Officers: An

Unexpected Finding." *International Journal of Offender Therapy and Comparative Criminology*, 35(1), 79.

Hess, Henry (1996, October 23). Former constable describes tampering. *The Globe and Mail*, p. A9.

Hobbes, Thomas (1651). *Leviathan*, reprinted by Bobbs-Merrill, 1958.

Horin, Adele. (1994, July 9). Why loyalty is a two-way street. *The Sydney Morning Herald*.

Janis, Irving L. (1968). "Group Dynamics under Conditions of External Danger." In: *Group Dynamics—Research and Theory*. Darwin Cartwright and Alvin Zander, eds. Harper and Row, Publishers, New York.

Josephson, Michael. "The Bell, the Book, the Candle." The Josephson Institute for Ethics, California.

Kant, Immanuel (1785). *Foundations of the Metaphysics of Morals*.

Kauffman, Kelsey (1988). *Prison Officers and Their World*. Harvard University Press, Cambridge, Massachusetts.

Klofas, John and Hans Toch (1982). "The Guard Subculture Myth." *Journal of Research in Crime and Delinquency*, July, 238–254.

Kohlberg, Lawrence (1984). *Essays on Moral Development, Vol. II: The Psychology of Moral Development*. Harper and Row, San Francisco.

Lamont, Leonie (1997, May 17). Force's hope for the future seeks a keen eye and a clean nose. *The Sydney Morning Herald*, p. 8.

Levy, Harold (1997, March 7). Courtroom 'tense' for detective in shooting. *The Toronto Star*, p. A6.

Lickona, Thomas (1976). "What Does Moral Psychology Have to Say?" In: *Moral Development and Behaviour: Theory, Research and Social Issues*. Lickona, Thomas, Gilbert Geis and Lawrence

Kohlberg, eds. Holt, Rinehart, Winston, New York.

Main, Eron (1996, October 1). Discretion no part of valour. *The Globe and Mail*, p. A16.

Martinson, Barbara, ed. (1987). *A Discourse on Ethics and the Corporate Workplace: Can Ethics Be Taught?* Working Paper No. 2. Corporate Council on the Liberal Arts, Working Paper Series, New York.

Mascoll, Philip (1996, November 21). Disciplinary muscle proposed for chief. *The Toronto Star*, p. A6.

Mill, John Stuart (1973). "Utilitarianism." In: *The Utilitarians*. Bentham, Jeremy and Mill, John Stuart, Academic Press, Doubleday.

Millar, Cal and Philip Mascoll (1995, March 17). Four officers suspended in probe. *The Toronto Star*, p. A6.

Mills, Theodore (1967). *The Sociology of Small Groups*. Prentice Hall, Englewood, New Jersey.

National Post (2000, May 5). One good cop takes his leave, p A21.

Noddings, Nel (1994). "Conversation as Moral Education." *Journal of Moral Education*, Vol. 23, No. 2, 107–118.

Paul, Richard (1992). *Critical Thinking*. (Revised Second Edition). Foundation for Critical Thinking, Sonoma State University, Rohnert Park, California.

Plato (1955). *The Republic of Plato*, translated by Francis Macdonald Cornford. Clarendon Press, London.

Plato (1954). "The Apology." In: *The Last Days of Socrates*, translated by Hugh Tredennick. Penguin Classics.

The Senate. Task Force on Ethics (1988). Progress Review: The University of Alberta. Alberta University, Edmonton.

The Senate. Task Force on Ethics (1985). The University of Alberta. Alberta University, Edmonton.

Sheehy, G. (1977). *Passages: Predictable Crises in Adult Life*. Bantam, New York.

Solomon, Robert (1990). *The Big Questions* (Third Edition). Harcourt Brace Jovanovich, New York.

Strike, Kenneth A., Emile J. Haller, and Jonas F. Soltis (1988). *The Ethics of School Administration*. Teachers College Press, New York.

Sydney Morning Herald (2002). "I was an outcast," says whistle-blower, p. 4 News.

Tafoya, William (1995). "Ethics and the Realities of Life: Surviving the Vortex." In Daryl Close and Nicholas Meier, eds. *Morality in Criminal Justice*. Wadsworth Publishing Company, San Francisco.

Toronto Star (2001, May 29). Unrepentant drunk driver utters a stunning last word, p. A4.

Weber, Max (1967). In H.H. Gerth and C. Wright Mills, eds. *Essays in Sociology*. New York and Oxford University Press, New York.

Index